Encyclopedia of Optical Fiber Technology: Progress in Research

Volume V

Encyclopedia of
Optical Fiber Technology:
Progress in Research
Volume V

Edited by **Marko Silver**

New York

Published by NY Research Press,
23 West, 55th Street, Suite 816,
New York, NY 10019, USA
www.nyresearchpress.com

Encyclopedia of Optical Fiber Technology: Progress in Research
Volume V
Edited by Marko Silver

International Standard Book Number: 978-1-63238-149-1 (Hardback)

Printed in the United States of America.

Contents

Preface

This book has been a concerted effort by a group of academicians, researchers and scientists, who have contributed their research works for the realization of the book. This book has materialized in the wake of emerging advancements and innovations in this field. Therefore, the need of the hour was to compile all the required researches and disseminate the knowledge to a broad spectrum of people comprising of students, researchers and specialists of the field.

This book presents up to date information and latest developments in optical fiber research. It discusses various topics such as new applications for optical fibers; and nonlinear and polarization effects in optical fibers. It presents detailed study of nonlinear effects in optical fibers in terms of experiments, applications and theoretical analysis. It further discusses fiber birefringence effects, chromatic dispersion and polarization dependent losses in optical fibers, spun fibers and polarization mode dispersion. This book includes contributions by eminent scientists and practitioners having vast knowledge and experience in the field of optics. The book provides latest developments in optical fiber research for the benefit of researchers, students and industrial users associated with the field of optical fiber technologies.

At the end of the preface, I would like to thank the authors for their brilliant chapters and the publisher for guiding us all-through the making of the book till its final stage. Also, I would like to thank my family for providing the support and encouragement throughout my academic career and research projects.

Editor

Part 1

Nonlinear Effects in Optical Fibers

Multimode Nonlinear Fibre Optics: Theory and Applications

Peter Horak and Francesco Poletti
University of Southampton
United Kingdom

1. Introduction

Optical fibres have been developed as an ideal medium for the delivery of optical pulses ever since their inception (Kao & Hockham, 1966). Much of that development has been focused on the transmission of low-energy pulses for communication purposes and thus fibres have been optimised for singlemode guidance with minimum propagation losses only limited by the intrinsic material absorption of silica glass of about 0.2dB/km in the near infrared part of the spectrum (Miya et al., 1979). The corresponding increase in accessible transmission length simultaneously started the interest in nonlinear fibre optics, for example with early work on the stimulated Raman effect (Stolen et al., 1972) and on optical solitons (Hasegawa & Tappert, 1973). Since the advent of fibre amplifiers (Mears et al., 1987), available fibre-coupled laser powers have been increasing dramatically and, in particular, fibre lasers now exceed kW levels in continuous wave (cw) operation (Jeong et al., 2004) and MW peak powers for pulses (Galvanauskas et al., 2007) in all-fibre systems. These developments are pushing the limits of current fibre technology, demanding fibres with larger mode areas and higher damage threshold. However, it is increasingly difficult to meet these requirements with fibres supporting one single optical mode and therefore often multiple modes are guided.

Non-fibre-based laser systems are capable of delivering even larger peak powers, for example commercial Ti:sapphire fs lasers now reach the GW regime. Such extreme powers cannot be transmitted in conventional glass fibres at all without destroying them (Gaeta, 2000), but there is a range of applications for such pulses coupled into hollow-core capillaries, such as pulse compression (Sartania et al., 1997) and high-harmonic generation (Rundquist et al., 1998). For typical experimental parameters, these capillaries act as optical waveguides for a large number of spatial modes and modal interactions contribute significantly to the system dynamics.

In order to design ever more efficient fibre lasers, to optimise pulse delivery and to control nonlinear applications in the high power regime, a thorough understanding of pulse propagation and nonlinear interactions in multimode fibres and waveguides is required. The conventional tools for modelling and investigating such systems are based on beam propagation methods (Okamoto, 2006). However, these are numerically expensive and provide little insight into the dependence of fundamental nonlinear processes on specific fibre properties, e.g., on transverse mode functions, dispersion and nonlinear mode coupling. For such an interpretation a multimode equivalent of the nonlinear Schrödinger equation, the standard and highly accurate method for describing singlemode nonlinear pulse propagation (Agrawal, 2001; Blow & Wood, 1989), is desirable.

In this chapter, we discuss the basics of such a multimode generalised nonlinear Schrödinger equation (Poletti & Horak, 2008), its simplification to experimentally relevant situations and a few select applications. We start by introducing and discussing the theoretical framework for fibres with $\chi^{(3)}$ nonlinearity in Sec. 2. The following sections are devoted to multimode nonlinear applications, presented in the order of increasing laser peak powers. A sample application in the multi-kW regime is supercontinuum generation, discussed in Sec. 3. Here we demonstrate how fibre mode symmetries and launching conditions affect intermodal power transfer and spectral broadening. For peak powers in the MW regime, self-focusing effects become significant and lead to strong mode coupling. The spatio-temporal evolution of pulses in this limit is the topic of Sec. 4. Finally, at GW peak power levels, optical pulses can only be delivered by propagation in gases. Still, intensities become so high that nonlinear effects related to ionisation must be taken into account. An extension of the multimode theory to include these extreme high power effects is presented in Sec. 5 and the significance of mode interaction is demonstrated by numerical examples pertaining to a recent experiment. Finally, we end this chapter with conclusions in Sec. 6.

2. The multimode generalised nonlinear Schrödinger equation

Pulse propagation in singlemode fibres is frequently modelled by a generalised nonlinear Schrödinger equation (NLSE) which describes the evolution of the electric field amplitude envelope of an optical pulse as it propagates along the fibre (Agrawal, 2001; Blow & Wood, 1989). This framework has been extremely successful in incorporating all linear and nonlinear effects usually encountered in fibres, such as second and higher order dispersion, Kerr and Raman nonlinearities and self-steepening, and its predictions have been corroborated by numerous experiments using conventional fibres, photonic crystal fibres and fibre tapers of different materials, as well as laser sources from the continuous wave regime down to few cycle pulses. Perhaps the most prominent application of the NLSE is in the description of supercontinuum generation where all the linear and nonlinear dispersion effects come together to induce spectacular spectral broadening of light, often over very short propagation distances (Dudley et al., 2006).

For very high power applications, as motivated above, a further extension of the NLSE is required to deal with the multimode aspects of large-mode area fibres. A very general multimode framework has been presented recently allowing for arbitrary mode numbers, polarisations, tight mode confinements and ultrashort pulses (Poletti & Horak, 2008). Here we describe a slightly simplified version that is more easily tractable and still is applicable to many realistic situations, e.g., for the description of high power applications as discussed in the later sections.

We start by considering a laser pulse propagating in a multimode fibre. The pulse can be written as the product of a carrier wave $\exp[i(\beta_0^{(0)}z - \omega_0 t)]$, where ω_0 is the carrier angular frequency and $\beta_0^{(0)}$ is its propagation constant in the fundamental fibre mode, and an envelope function $\mathbf{E}(\mathbf{x}, t)$ in space and time. Note that throughout this chapter we adopt the notation that vectorial quantities are written in bold face and $\mathbf{x} = (x, y, z)$. For convenience, we assume $\mathbf{E}(\mathbf{x}, t)$ to be complex-valued, so that it includes the envelope phase as well as the amplitude, and we consider the pulse evolution in a reference frame moving with the group velocity of the fundamental mode, so that in the absence of dispersion a pulse would stay centred at time

$t = 0$ throughout its propagation. Finally, we use units such that $|\mathbf{E}(\mathbf{x}, t)|^2$ is the field intensity in W/m^2. The envelope function can then be expanded into a superposition of individual modes $p = 0, 1, 2, ...$, each represented by a discrete transverse fibre mode profile $\mathbf{F}_p(x, y)$ and a modal envelope $A_p(z, t)$, as

$$\mathbf{E}(\mathbf{x}, t) = \sum_p \frac{\mathbf{F}_p(x, y)}{\left[\int dx\, dy\, |\mathbf{F}_p|^2\right]^{1/2}} A_p(z, t). \qquad (1)$$

Note that $|A_p(z, t)|^2$ gives the instantaneous power propagating in mode p in units of W, and that a simplified normalisation has been used compared to a more rigorous previous formulation (Poletti & Horak, 2008). The accuracy of this approximation improves as the fibre core size is increased and the core-cladding index contrast is decreased, leading to guided modes with an increasingly negligible longitudinal component of polarisation.

The multimode generalised nonlinear Schrödinger equation (MM-NLSE) is then given by the following set of coupled equations to describe the dynamics of the mode envelopes,

$$\frac{\partial A_p}{\partial z} = \mathcal{D}\{A_p\}$$
$$+ i \frac{n_2 \omega_0}{c} \left(1 + \frac{i}{\omega_0} \frac{\partial}{\partial t}\right) \sum_{l,m,n} \left\{(1 - f_R) S^K_{plmn} A_l A_m A^*_n + f_R S^R_{plmn} A_l [h * (A_m A^*_n)]\right\}. \qquad (2)$$

The following approximations have been applied here: (i) we have assumed that the Raman response and the pulse envelope functions vary slowly on the time scale of a single cycle of the carrier wave, so that we can neglect a rapidly oscillating term, and (ii) an additional term related to the frequency dependence of the mode functions has been omitted, assuming the variation of $S^{K,R}_{plmn}$ is slow compared to the $1/\omega_0$ self-steepening term. In Eq. (2),

$$\mathcal{D}\{A_p\} = i(\beta_0^{(p)} - \Re[\beta_0^{(0)}]) A_p - (\beta_1^{(p)} - \Re[\beta_1^{(0)}]) \frac{\partial A_p}{\partial t} + i \sum_{n \geq 2} \frac{\beta_n^{(p)}}{n!} \left(i \frac{\partial}{\partial t}\right)^n A_p \qquad (3)$$

yields the effects of dispersion of mode p with coefficients $\beta_n^{(p)} = \partial^n \beta^{(p)} / \partial \omega^n$. Here we allow for complex values of the modal propagation constants $\beta^{(p)}$ where the imaginary part describes mode and wavelength dependent losses; $\Re[..]$ denotes the real part only. The second line of (2) represents the effects of optical nonlinearity with a nonlinear refractive index n_2. The term $\propto \partial/\partial t$ describes self-steepening and the two terms within the sum describe Kerr and Raman nonlinearities. The Raman term contributes with a fraction f_R to the overall nonlinearity ($f_R = 0.18$ for silica glass fibres) and contains the Raman mode overlap factors

$$S^R_{plmn} = \frac{\int dx\, dy\, [\mathbf{F}^*_p \cdot \mathbf{F}_l][\mathbf{F}_m \cdot \mathbf{F}^*_n]}{\left[\int dx\, dy\, |\mathbf{F}_p|^2 \int dx\, dy\, |\mathbf{F}_l|^2 \int dx\, dy\, |\mathbf{F}_m|^2 \int dx\, dy\, |\mathbf{F}_n|^2\right]^{1/2}} \qquad (4)$$

as well as a convolution of the time dependent Raman response function $h(t)$ with two mode amplitudes

$$[h * (A_m A^*_n)](z, t) = \int d\tau\, h(\tau) A_m(z, t - \tau) A^*_n(z, t - \tau). \qquad (5)$$

The mode overlap factors responsible for the instantaneous Kerr effect are given by

$$S_{plmn}^{K} = \frac{2}{3}S_{plmn}^{R} + \frac{1}{3}\frac{\int dx\,dy\,\left[\mathbf{F}_{p}^{*}\cdot\mathbf{F}_{n}^{*}\right]\left[\mathbf{F}_{m}\cdot\mathbf{F}_{l}\right]}{\left[\int dx\,dy\,|\mathbf{F}_{p}|^{2}\int dx\,dy\,|\mathbf{F}_{l}|^{2}\int dx\,dy\,|\mathbf{F}_{m}|^{2}\int dx\,dy\,|\mathbf{F}_{n}|^{2}\right]^{1/2}}. \tag{6}$$

Numerically, the mode functions of all the modes involved in the nonlinear effects under consideration are first evaluated at ω_0 and a table of overlap integrals is calculated. The number of modes and overlap integrals can be greatly reduced based on mode symmetry arguments (Poletti & Horak, 2008); all the applications discussed in the following will employ such reduced sets of modes. Next, the dispersion curves for these modes are calculated. Finally, the system of equations (2) is integrated numerically using a standard symmetrised split-step Fourier method (Agrawal, 2001), where adaptive step size control is implemented by propagating the nonlinear operator using a Runge-Kutta-Fehlberg method (Press et al., 2006). In order to avoid numerical artifacts, we also found it necessary to further limit the maximum step size to a fraction of the shortest beat length between all the modes considered. The accuracy and convergence of the results is further checked by running multiple simulations with increasingly small longitudinal step sizes.

The framework presented above still allows for modes of arbitrary polarisation. In most practical situations, however, one is interested in a subset of modes representing only a specific polarisation state which is determined by the pump laser. The two most common cases are briefly discussed in the following.

2.1 Circular polarisation

Under the weak guiding condition, modes fall into groups of LP_{mn} modes containing either two ($m = 0$) or four ($m > 0$) degenerate modes. Within each group, the modes can be combined into modes that are either σ_+ or σ_- circularly polarised at every point in the fibre. If the light launched into the fibre is, for example, σ_+ polarised, the form of the overlap integrals (4) and (6) guarantees that no light is coupled into the σ_- polarised modes during propagation and those modes can therefore be eliminated entirely from the model. It is worth emphasising that this is an *exact result* within the weak guiding limit. Using the properties of circular polarisation vectors, the overlap integrals are then simplified to

$$S_{plmn}^{R} = \frac{\int dx\,dy\,F_{p}F_{l}F_{m}F_{n}}{\left[\int dx\,dy\,F_{p}^{2}\int dx\,dy\,F_{l}^{2}\int dx\,dy\,F_{m}^{2}\int dx\,dy\,F_{n}^{2}\right]^{1/2}},$$

$$S_{plmn}^{K} = \frac{2}{3}S_{plmn}^{R}, \tag{7}$$

where the mode functions have been written as $\mathbf{F}_{p} = \mathbf{e}_{+}F_{p}$ for σ_{+} polarised modes with real-valued scalar mode functions F_{p}.

2.2 Linear polarisation

The situation is slightly more complicated in the case of linearly polarised pump light. In this case, nonlinear coupling between orthogonal polarisation modes is in principle allowed, leading to, for example, birefringent phase matching and vector modulation instability (Agrawal, 2001; Dupriez et al., 2007). However, for many practical situations where modes can be described as LP_{mn} modes, if linearly polarised light is launched into the fibre, nonlinear coupling to orthogonal polarisation states is effectively so small that most of the pulse energy remains in its original polarisation throughout the entire pulse propagation. This allows halving the number of modes to be considered in the model with significant computational advantage, and a simpler definition of the overlap factors (4) and (6). There are several important practical situations where this approximation can be acceptable:

(i) For degenerate modes (no birefringence), the overlap factor (6) for four-wave mixing (FWM) between modes of parallel polarisation is three times larger than that for orthogonal polarisation. Since the dispersion properties, and therefore the phase matching conditions, are the same, nonlinear gain is much higher for the same polarisation and thus will dominate the dynamics.

(ii) For few-moded fibres power transfer to orthogonal modes by FWM can be negligible if either the phase matching condition cannot be fulfilled at all, or if the phase matching condition is achieved only for widely separated wavelength bands where the difference in group velocities limits the effective interaction length due to walk-off effects.

In these situations one can therefore use an approximate theoretical description of pulse propagation by restricting the MM-NLSE to the LP_{mn} modes of the fibre with the same linear polarisation everywhere. Assuming real-valued x-polarised mode functions $\mathbf{F}_p = \mathbf{e}_x F_p$, the overlap integrals then reduce to

$$S_{plmn}^R = S_{plmn}^K = \frac{\int dx\, dy\, F_p F_l F_m F_n}{\left[\int dx\, dy\, F_p^2 \int dx\, dy\, F_l^2 \int dx\, dy\, F_m^2 \int dx\, dy\, F_n^2 \right]^{1/2}}. \tag{8}$$

A further simplification is also sometimes possible. If linearly polarised light is predominantly launched in an LP_{0n} mode, power transfer into LP_{mn} modes with $m > 0$ can only be initiated by spontaneous FWM processes. By contrast, other LP_{0n} modes of the same polarisation can be excited by stimulated processes, see Sec. 3.1. Thus, if the dominant processes within the pulse propagation are stimulated ones, e.g., in the regime of high powers and relatively short propagation distances, the study can be effectively restricted to LP_{0n} modes with the same polarisation.

3. Supercontinuum generation in multimode fibres

One of the first applications where the MM-NLSE presented in the previous section can provide deep insights is that of supercontinuum (SC) generation in multimode fibres. As already mentioned, the complex dynamic underlying SC generation in *singlemode* fibres is by now well understood. Octave spanning SC in suitably designed fibres arises as a combination of various nonlinear phenomena, including soliton compression and fission, modulation instability, parametric processes, intrapulse Raman scattering, self phase modulation (SPM) and cross phase modulation (XPM) (Dudley et al., 2006). As the fibre diameter is increased though, as required for example to increase the SC power spectral density without destroying the fibre, the fibre starts to support multiple modes. Previous theoretical models were usually restricted to two polarisation modes of a birefringent fibre (Agrawal, 2001; Coen et al., 2002; Lehtonen et al., 2003; Martins et al., 2007) or included a maximum of two spatially distinct modes (Dudley et al., 2002; Lesvigne et al., 2007; Tonello et al., 2006). Using the full MM-NLSE, however, fibres with arbitrary modal contents can be studied, for which a rich new list of *intermodal* nonlinear phenomena emerges, causing the transfer of nonlinear phase and/or power between selected combinations of modes (Poletti & Horak, 2009).

In this section, using simulations of a specific few-moded fibre as an illustrative example, we will discuss how modal symmetries and launch conditions can have a drastic influence on intermodal power transfer dynamics. For pump peak powers in the range of tens to hundreds of kW, if the nonlinear length of the pump pulses is shorter than the walk-off length between the modes involved, significant power transfer into high-order modes with the appropriate symmetry can occur, which can be beneficial, for example, to further extend the SC spectrum to shorter wavelengths. Even if conditions for significant intermodal power transfer are not met, it is found that intermodal XPM can still play a significant role in the SC dynamics by broadening the spectrum of modes which would not otherwise present a significant spectral broadening if pumped on their own.

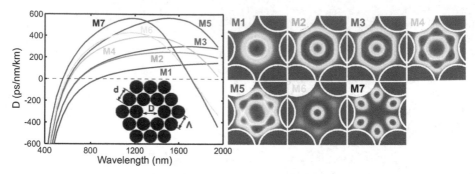

Fig. 1. GVD curves and transverse mode functions, calculated at 850nm, of the 7 circularly polarised modes guided in a HF with $\Lambda = 2.7\mu$m and $d = 2.5\mu$m .

To discuss the intermodal nonlinear dynamics leading to SC generation we focus on a *moderately* multimoded holey fibre (HF) consisting of two rings of large circular air holes with pitch $\Lambda = 2.7\mu$m and relative hole size $d/\Lambda = 0.93$, surrounding a solid core with a diameter of a few optical wavelengths ($D = 2\Lambda - d = 2.9\mu$m), see Fig. 1. From 400nm to 2000nm the fibre supports 14 modes with effective areas ranging between 3.6 and 6.1μm^2. To reduce the computational time it is possible to combine these modes into 7 pairs of circularly polarised modes and to exploit the forbidden power exchange between modes with opposite circular polarisation (see Sec. 2.1), only to focus on the 7 right-handed circularly polarised modes M1, M2,..., M7 shown in Fig. 1. The group velocity dispersion (GVD) curves of these modes are significantly different from each other, with a first zero dispersion wavelength (ZDW) ranging from $\lambda_7 = 550$nm for M7 to $\lambda_1 = 860$nm for M1.

3.1 Effect of modal symmetries and launch conditions on intermodal power transfer
Equation (2) shows that the transfer of power between modes is mediated by FWM terms of the form $S^K_{plmn}A_lA_mA_n^*$, with $l, m \neq n$. If only a single mode l is initially excited with a narrow spectral line, the strongest power transfer to mode p and therefore the first to be observed in the nonlinear process is the one controlled by *degenerate* FWM terms of the form $S^K_{plln}A_lA_lA_n^*$. If both modes p and n are initially empty, power transfer starts with a spontaneous FWM process and is therefore slow. If one of the generated photons is however returned into the pump l by *stimulated emission*, the process becomes much faster and tends to dominate the nonlinear dynamics in the limit of high-power pulse propagation over short distances. Interestingly, these $S^K_{plll}A_lA_lA_l^*$ processes produce automatic phase-locking of mode p to the pump mode l, similarly to what happens in non-phase matched second and third harmonic generation processes (Roppo et al., 2007). However, processes S^K_{plll} require (i) that modes p and l belong to the same symmetry class, and (ii) that they present a large overlap. For the HF

under investigation these conditions are only fulfilled for the two LP_{0n} modes M1 and M6, and therefore one would expect significant power transfer only between them.

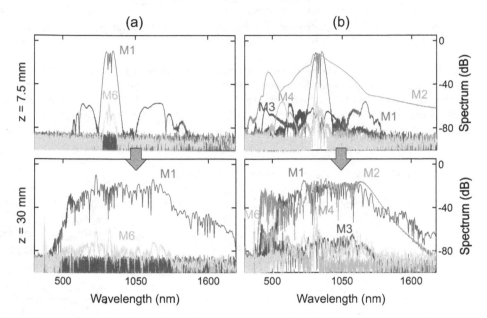

Fig. 2. Simulations of multimode nonlinear propagation in the HF of Fig. 1 after 7.5mm (top row) and 30mm (bottom row), for a 100fs sech-shaped pump centred at 850nm. (a) Only M1 is excited and (b) both M1 and M2 are excited with a 50kW peak power pulse.

This expected behaviour is indeed confirmed by the numerical simulation shown in Fig. 2(a), where a hyperbolic secant pump pulse with temporal profile $A_p(0,t) = \sqrt{P_0} \operatorname{sech}(t/T_0)$ with $T_0 = 100$fs (full width at half maximum 176fs) and centred at $\lambda_p = 850$nm is launched into M1 only and propagated through 30mm of the HF. Here the pulse peak power P_0 is set to 50kW, corresponding to a 10nJ pulse and, for mode M1, to a soliton of order $N = 166$. As one would expect from single mode SC theory (Dudley et al., 2006), besides SPM-induced spectral broadening, such a high-N pulse develops sidebands which grow spontaneously from noise, through an initial modulation instability (MI) process. The characteristic distance of this phenomenon $L_{MI} \sim 16L_{NL} = 16\lambda/(3\pi n_2 S_{1111}^K P_0) = 6.9$mm correlates well with the simulation results. As expected, of all the other 6 modes only M6 is significantly amplified at wavelengths around λ_p, and subsequently develops a wide spectral expansion and an isolated peak at 360nm. Further analysis of spectrograms and phase matching conditions indicates that this peak is a dispersive wave in M6, phase matched to a soliton in M1 and slowly shifting to shorter wavelengths as the soliton red-shifts due to the effect of intrapulse Raman nonlinearity. Under these launching conditions the study can thus be restricted to the LP_{0n} modes of the fibre without loss of accuracy. Simulations also show that if either M2, M3, M4, M5 or M7 are selectively launched, no power is transferred to any of the other modes, and each of them evolves as in the single mode case.

When two or more modes contain a significant amount of power, they can all act as pumps for weaker modes. Moreover, if these modes belong to different symmetry classes, additional

FWM terms come into play, giving rise to a much richer phenomenology. As an example, Fig. 2(b) shows what happens when both M1 and M2 are simultaneously excited with a $P_0 = 50\text{kW}$ sech pulse. This pulse corresponds to an $N = 27$ soliton for M2, due to its much larger value of $\beta_2^{(2)}$ at the pump wavelength. As a result, the SC generated in M2 has a more temporally coherent nature, as it originates from soliton compression and fission mechanisms (the fission length $L_{\text{fiss}} = N \cdot L_{NL}$ is around 16mm). Due to a shorter ZDW than M1, the final SC in M2 also extends to much shorter wavelengths than the one in M1 (400nm versus 550nm, respectively), which can be one of the benefits of using multimode fibres for SC generation. Moreover, in addition to M6, also M3 and M4 are amplified from noise, generating a complex output spectrum, where the final relative magnitude of different modes is a strong function of wavelength. This is reminiscent of early experimental results (Delmonte et al., 2006; Price et al., 2003).

3.2 Non-phase matched permanent intermodal power transfer

To understand the complex dynamics of intermodal power transfer it is useful to refer to the approximate analytical theory of cw pumped parametric processes, which neglects the effects of GVD and pulse walk-off but still provides a valid reference (Stolen & Bjorkholm, 1982). Within this framework, parametric gain leading to exponential signal amplification requires the propagation constant mismatch $\Delta\beta_{plmn} = \beta^{(l)}(\omega_l) + \beta^{(m)}(\omega_m) - \beta^{(p)}(\omega_p) - \beta^{(n)}(\omega_n)$ to be smaller than a few times the *average* inverse nonlinear length $1/L_{NL} = \overline{\gamma}P_0$.

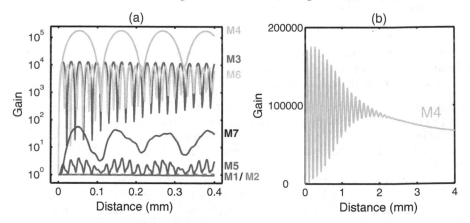

Fig. 3. (a) Dynamic gain evolution for each individual mode when M1 and M2 are simultaneously excited at launch as in Fig. 2(b), showing the oscillatory behaviour typical of non-phase matched parametric processes. (b) Permanent power transfer to M4 despite the lack of parametric phase matching due to walk-off between the pumps in M1 and M2 and the signal in M4.

For multimode processes, an estimate of $\overline{\gamma}$ can be obtained by averaging all the intermodal nonlinearities $\gamma_{plmn} = \frac{3\pi n_2}{\lambda} S_{plmn}^K$ which contribute to SPM and XPM between the relevant modes. However, in most practical situations involving SC generation in highly nonlinear multimode fibres, $\Delta\beta_{plmn} \gg \overline{\gamma}P_0$ for all the relevant FWM processes considered. Thus, no parametric gain is typically observed and each FWM term leads to an oscillatory power exchange between modes, as shown by the dynamic gain curves of high order modes when only M1 and M2 are initially pumped, reported in Fig. 3(a). The oscillation periods are given by the beat lengths $L_b \sim 2\pi/|\Delta\beta|$. For example, for the process leading to amplification of M6, $\Delta\beta_{6111} = 4.1 \cdot 10^5$ m^{-1}, corresponding to a value of $L_b = 15.3\mu$m in agreement with the simulation. For modes amplified by a cascade of intermodal FWM processes, such as M5 and M7 in the example, the signature of multiple beating frequencies can be clearly observed.

Despite the non-phase matched nature of most FWM processes, simulations show that after long enough propagation some power is *permanently* transferred into the weaker modes. This is shown, for example, in Fig. 3(b) extending the propagation distance of M4 from 0.4mm to 4mm. A more detailed analysis excluding XPM and Raman effects found this behaviour to be uniquely caused by the temporal walk-off between the pulses involved. The typical length scale of this permanent power transfer is therefore of the order of the walk-off length of all the pulses involved, given by $L_W^{pq} = T_0/|1/v_g^{(p)} - 1/v_g^{(q)}| = T_0/|\beta_1^{(p)} - \beta_1^{(q)}|$ for modes p and q. For the example in Fig. 3(b), $L_W^{12} = 3$mm, $L_W^{24} = 2.4$mm and $L_W^{14} = 1.3$mm, which correlate well with the simulation.

In conclusion, nonlinear intermodal power transfer is governed by two length scales, a beat length leading to fast initial power oscillations and a walk-off length leading to permanent power transfer. In order to observe in practice intermodal nonlinear effects, the nonlinear length of the pump pulses must be shorter than the walk-off length, i.e., high peak powers are required. Otherwise, nominally multimode fibres can exhibit the same nonlinear behaviour as singlemode ones. Scaling a fixed fibre structure to larger core sizes allows for larger power throughput, but at the same time longer beat and walk-off lengths lead to much stronger mode coupling, and significant amounts of power can be transferred into higher order modes. In this case, as shown in Fig. 2, higher order modes may also serve to extend the SC spectral extension to much shorter wavelengths.

3.3 Effect of intermodal cross phase modulation

Intermodal power transfer mediated by FWM terms, which can permanently exchange power between modes even in the absence of proper phase matching, is not the only intermodal nonlinear effect which can occur in a multimode fibre. Intermodal XPM can also play a role in significantly broadening the spectrum of a mode which would not undergo a significant spectral expansion if propagated on its own (Chaipiboonwong et al., 2007; Schreiber et al., 2005).

To illustrate this phenomenon, we simulate the propagation of a pulse launched in M1 and/or M2 at 725nm, where M1 is in the normal dispersion region and M2 is in the anomalous region. In order to observe significant spectral expansion and intermodal effects within the distance where the pulses are temporally overlapped, we increase the input power up to a value of $P_0 = 500$kW, close to the estimated fibre damage threshold.

Figs. 4(a) and (b) show that when M1 is individually launched, only some SPM-based spectral expansion is visible, whereas if only M2 is launched, a wide MI-based SC develops. On the other hand, if the same input pulse is launched simultaneously in *both* modes as in Fig. 4(c), a much wider output spectrum is developed also in M1. Under these operating conditions the intermodal power transfer is negligible, as confirmed by nearly identical spectral results obtained when all S_{plmn}^K and S_{plmn}^R coefficients responsible for intermodal FWM are set to zero. Therefore, the increased spectral expansion in M1 must be generated by intermodal XPM effects alone. This is indeed confirmed by the simulation in Fig. 4(d), showing that when all intermodal XPM effects are artificially switched off, M1 and M2 produce a very similar spectrum to that of their individual propagation.

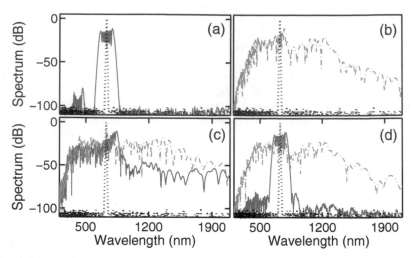

Fig. 4. Spectral output after 2mm propagation in the HF of Fig. 1 of a $T_0 = 100$fs and $P_0 = 500$kW sech pulse centred at 725nm and launched in: (a) M1 only (blue, solid line); (b) M2 only (green, dashed-dot line); (c) both M1 and M2, and (d) both M1 and M2 when all intermodal XPM coefficients are artificially set to zero. The input pulse is shown as a black dotted line.

4. Self-focusing in optical fibres in a modal picture

For laser powers larger than discussed in the previous section and into the MW regime, the nonlinear refractive index induced in the glass by the laser may become strong enough to introduce significant spatial reshaping of the beam in the transverse direction. The refractive index of a material is given by $n_0 + n_2 I$, including both the linear, n_0, and nonlinear term, n_2, and where I is the position-dependent intensity of the laser. Thus, if the beam has a Gaussian-like transverse profile and the optical Kerr nonlinearity n_2 is positive, as is the case in most of the commonly used transparent materials, the induced nonlinear refractive index is maximum at the centre of the beam and decreases towards the pulse edges. Therefore, the induced index profile forms a focusing lens, acting back on the laser beam itself. This effect is known as *self-focusing* and has been studied extensively in bulk materials for nearly 50 years (Askaryan, 1962; Chiao et al., 1964). For input powers P below a critical power P_{crit}, self-focusing is finally overcome by the beam divergence. In the case of $P > P_{crit}$, however, the pulse undergoes catastrophic collapse leading to permanent damage of the material (Gaeta, 2000). The critical power is given by

$$P_{crit} = 1.86 \frac{\lambda^2}{4\pi n_0 n_2}, \tag{9}$$

where the numerical factor slightly depends on the beam profile in a bulk material (Fibich & Gaeta, 2000). Numerically, self-focusing in bulk media is most commonly modelled by slowly-varying envelope models or, more accurately, by a nonlinear envelope equation (NEE) describing the dynamics of the transverse beam profile $\Phi(\mathbf{x}, t)$ (Brabec & Krausz, 1997;

Ranka & Gaeta, 1999),

$$\frac{\partial}{\partial z}\Phi = \mathcal{D}_{mat}\{\Phi\} + \frac{i}{2\beta_0}\left(1 + \frac{i}{\omega_0}\frac{\partial}{\partial t}\right)^{-1}\nabla_\perp^2\Phi + i\frac{n_2 n_0 \omega_0}{2\pi}\left(1 + \frac{i}{\omega_0}\frac{\partial}{\partial t}\right)|\Phi|^2\Phi, \quad (10)$$

where $\mathcal{D}_{mat}\{\Phi\}$ is a dispersion term similar to (3) describing the effect of material dispersion and $\nabla_\perp^2$ is the transverse Laplace operator. The NEE incorporates many features similar to the MM-NLSE (2), e.g., higher order dispersion, Kerr nonlinearity and self-steepening terms. However, even in the presence of rotational symmetry, the envelope function Φ is a two-dimensional object (radial and temporal coordinate), in contrast to the MM-NLSE which only uses a finite number of one-dimensional (temporal) envelope functions to describe the same situation. If the number of modes is small, the MM-NLSE is thus computationally significantly more efficient, both in terms of reduced memory requirements and faster dynamics simulation.

It is now well established that the same process of self-focusing occurs in optical waveguides and fibres and that the same power threshold for catastrophic collapse applies (Farrow et al., 2006; Gaeta, 2000). However, for powers below P_{crit} the observed light propagation behaviour is qualitatively different from that observed in bulk media, since here the light is additionally bound by total internal reflection at the core-cladding interface, which can lead to additional spatial and temporal interference and dispersion effects, such as periodic oscillations of the beam profile or catastrophic pulse collapse even when the launched peak power is below the critical value. In this section we will discuss these effects within the framework of the MM-NLSE, which leads to an easy understanding of fibre-based self-focusing within a modal picture (Horak & Poletti, 2009; Milosevic et al., 2000). Such an interpretation is particularly useful in the context of high-power fibre lasers, which now achieve peak powers close to the critical power with pulse lengths approaching the nanosecond regime (Galvanauskas et al., 2007).

4.1 Continuous wave limit
We start our discussion with the case of cw propagation, which in practice is also a good approximation to the behaviour of long pulses (ps to ns regime) near the pulse peak, and use the MM-NLSE restricted to the linearly polarised LP_{0n} modes, as discussed in Sec. 2.2. The MM-NLSE thus reduces to

$$\frac{\partial A_p}{\partial z} = i(\beta_0^{(p)} - \beta_0^{(0)})A_p + i\frac{n_2\omega_0}{c}\sum_{l,m,n}S_{plmn}^K A_l A_m A_n^* \quad (11)$$

with S_{plmn}^K given by (8). Specifically, we assume propagation in a short piece of a step-index fibre with a pure silica core of $40\mu m$ diameter and a refractive index step of 0.02 between core and cladding. This fibre is similar to photonic crystal large-mode area fibres which are commercially available, where the index step has been increased such that the fibre supports eight LP_{0n} modes. The zero-dispersion wavelength of this fibre is at $1.26\mu m$, and we assume a pump laser operating at 1300nm wavelength. The critical power (9) for silica at this wavelength is $P_{crit} = 5.9$MW. Note that at this power level pulses up to approximately 100ps length can be transmitted through the fibre without fibre damage (Stuart et al., 1996). Figure 5 shows the dynamics of light propagation along this fibre when cw light is launched into the fundamental LP_{01} mode with a power of $0.7P_{crit}$=4.84MW. The curves in Fig. 5(a) show the power $|A_p|^2$ in the lowest order modes obtained by solving Eq. (11). Power

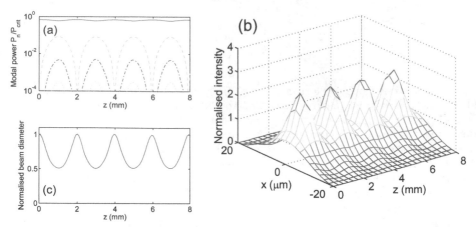

Fig. 5. Propagation of cw laser light at $1.3\mu m$ wavelength through a multimode silica step-index fibre with $40\mu m$ core diameter and core-cladding index difference of 0.02. The launched power is $0.7P_{crit}$=4.84MW in the fundamental LP_{01} mode. (a) Power in the lowest four fibre modes versus propagation distance. (b) 2D (transverse and longitudinal) spatial intensity profile of the beam. (c) Dynamics of the transverse beam width (FWHM), normalised to the width of the fundamental fibre mode.

from the fundamental mode is quickly transferred over sub-mm propagation distances into higher order modes by FWM processes, most prominently by induced FWM involving three pump photons as described by terms of the form $\partial A_p/\partial z \propto iA_0^2A_0^*$, see Sec. 3.1. However, because of the phase mismatch $\beta_0^{(p)} - \beta_0^{(0)}$ between the fundamental mode and the higher order modes the initial FWM gain is reversed after a certain propagation distance (about 1mm for the chosen parameters) and power is coherently transferred back into the pump from the higher order modes. This process is repeated subsequently leading to a periodic exchange of power between modes. The phase mismatch increases for increasing mode order and thus the maximum transferred power decreases.

In Fig. 5(b) we depict the corresponding 2D beam intensity $|\mathbf{E}(x,z)|^2$ calculated by summing the modal contributions (1), normalised to the maximum field $|\mathbf{E}(0,0)|^2$ at the fibre input. The field experiences significant periodic enhancement on the beam axis at positions where large fractions of the total power propagate inside higher order LP_{0n} modes. At these positions of enhanced intensity, the full width at half maximum (FWHM) of the beam profile is strongly reduced, as shown in Fig. 5(c). The intermodal FWM processes together with the modal phase mismatch are therefore responsible for periodic beam self-focusing and defocusing in a fibre. This complements the standard interpretation of self-focusing in a bulk medium using Gaussian beam propagation, which describes the same phenomenon as focusing by a Kerr-induced lensing effect, followed by beam divergence and subsequent total internal reflection at the core-cladding interface. We finally note that a stationary solution can be obtained for the cw MM-NLSE in which the modal amplitudes and phases are locked in such a way that no oscillations occur. In the bulk interpretation this corresponds to the situation where nonlinear focusing and diffraction are perfectly balanced, thereby generating a stationary spatial soliton.

It may seem that this modal description of self-focusing is only possible in multimode fibres but breaks down in singlemode fibres, for example in large-mode area photonic crystal fibres designed for endlessly single mode operation (Mortensen et al., 2003). However, in this case the role of the higher order bound modes of a multimode fibre is taken over by the cladding modes, and it is the FWM-induced power exchange between the guided mode of a singlemode fibre and its cladding modes which provides a modal interpretation of self-focusing.

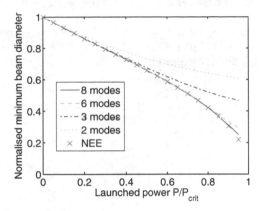

Fig. 6. Minimum beam diameter during the first period of self-focusing oscillation under cw pumping vs pump power for the same fibre parameters as in Fig. 5. The curves correspond to MM-NLSE simulations involving the lowest 2, 3, 6 modes only, and all 8 LP_{0n} modes (from top to bottom). The crosses indicate simulation results using the nonlinear envelope equation.

Using only a finite number of modes in the simulation of the MM-NLSE necessarily limits the transverse spatial resolution that can be achieved by this method. For example, the LP_{0n} mode function exhibits n maxima and $n-1$ zeros along the radial direction within the fibre core region. With simulations using n different modes one can therefore expect a maximum resolution of the order of R/n where R is the core radius. Simulations with pump powers approaching the critical power P_{crit} will thus require a larger number of modes in order to correctly describe the increasingly small minimum beam diameter. We investigate this behaviour in Fig. 6. Here we show the minimum beam diameter achieved during the first period of self-focusing and diffraction, i.e., at approximately 1mm of propagation for the parameters of Fig. 5, when the MM-NLSE is restricted to different numbers of modes. For clarity, the beam diameter is normalised to the diameter of the launched beam (LP_{01} mode). We observe that simulations with 2, 3, and 6 modes are accurate up to pump powers of approximately $0.2P_{crit}$, $0.4P_{crit}$, and $0.8P_{crit}$, respectively, compared to simulations involving all 8 bound fibre modes of this sample fibre. For comparison, we also show the results of the NEE beam propagation method (10). This confirms the accuracy of the MM-NLSE with 8 modes up to $0.95P_{crit}$ corresponding to a nearly five-fold spatial compression of the beam.

For the simulations shown in Fig. 6 we used the same 4th-5th order Runge-Kutta integration method with adaptive step size control (MATLAB R2010b by MathWorks, Inc.) for both the MM-NLSE and the NEE. Each data point required approximately 0.9s of CPU time on a standard desktop computer with the 8-mode MM-NLSE and <0.2s with 6 modes. In contrast, the corresponding NEE simulations with 1024 radial grid points required 101s, that is, two to three orders of magnitude slower than the MM-NLSE.

4.2 Short pulse propagation

Next, we consider the propagation of short pulses in the regime of peak powers close to the critical power, where in addition to transverse spatial effects the pulse may exhibit complex temporal dynamics related to intermodal and intramodal dispersion, self-steepening and nonlinear effects. As an example we consider sech-shaped pulses with a temporal FWHM of 100fs launched with a peak power of $0.8P_{crit}$ into the fundamental mode of the multimode fibre considered above. The pump wavelength is again set to 1.3μm. The simulations discussed in the following used a 6-mode MM-NLSE with 2048 temporal grid points solved with a split-step Fourier method (Poletti & Horak, 2008; 2009).

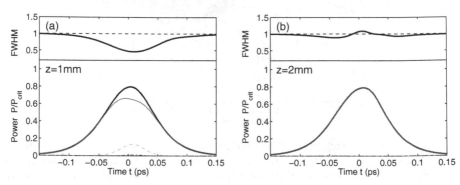

Fig. 7. Propagation of a 100fs sech-shaped pulse with $0.8P_{crit}$ peak power at 1.3μm wavelength with the same fibre parameters as in Fig. 5 after (a) 1mm and (b) 2mm of propagation. The bottom part of the figure shows the overall temporal pulse profile (thick solid line) as well as its contributions from the fundamental mode (thin solid), first (dashed) and second (dash-dotted) higher order modes. The top part of the figure shows the spatial FWHM beam diameter along the pulse, normalised to the FWHM of the fundamental mode.

The initial dynamics of the pulse propagation are shown in Fig. 7. After 1mm of propagation, Fig. 7(a), a significant amount of power has been transferred from the fundamental mode into the higher order modes, leading to a transverse beam focusing to approximately 40% of the input beam width. The transverse beam size depends on the pulse power and thus varies along the pulse shape: the beam diameter is smallest near the temporal peak of the pulse, but remains unchanged in the trailing and leading edges where the power is low. Propagating further to 2mm, Fig. 7(b), most of the power has been converted back into the fundamental mode, similar to the cw case of Fig. 5. However, the transfer is not complete and is not uniform along the pulse. This is related to the walk-off of the higher order modes because of intermodal dispersion as well as a slight dependence of the beam oscillation period on power. Therefore, the spatial FWHM of the beam at 2mm propagation length is below that of the fundamental mode in some parts of the pulse while it exceeds it in other parts.

Continuing the propagation of Fig. 7, the spatial beam variations persist, but the deviations from a simple oscillation become more prominent. This is shown clearly in Fig. 8(a) in the beam properties after 7mm of propagation. At this point the initial sech-shaped temporal profile has steepened on the trailing edge and an ultrashort pulse peak is forming due to the interference of the modal contributions. In particular, the first high order mode exhibits a similar power level as the fundamental mode. Simultaneously, the beam diameter is strongly reduced. At 7.4mm of propagation, Fig. 8(b), this peak has narrowed further and reaches

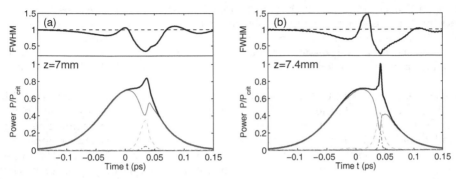

Fig. 8. Continuation of the pulse propagation of Fig. 7 to (a) 7mm and (b) 7.4mm of fibre length exhibiting simultaneous spatial and temporal collapse.

the critical power for catastrophic collapse while the beam diameter has reduced to 20% of the fundamental mode. For even longer propagation lengths the simulations show the pulse breaking up into many ultrashort high-intensity parts around this initial instability, however the MM-NLSE with 6 modes becomes invalid at this point due to its limited spatial and temporal resolution. Simulations with the MM-NLSE restricted to the fundamental mode reveal only a very small amount of pulse reshaping due to self-steepening over this propagation distance (a shift of the pulse peak by about 10fs) and exhibit none of the complex dynamics seen in Fig. 8. We therefore conclude that the simultaneous spatial and temporal collapse of the pulse observed here is a pure multimode effect, driven by FWM-based power exchange together with modal dispersion and self-steepening, in agreement with investigations based on beam propagation methods (Zharova et al., 2006).

5. Multimode effects in gas-filled waveguides

As discussed above, the peak power that can be transmitted in optical fibres is limited by the critical power for self-focusing and catastrophic collapse to levels of a few MW. According to Eq. (9), for a fixed laser wavelength P_{crit} only depends on the material linear and nonlinear refractive index. In general, the linear refractive index does not vary much across transparent media, between 1 for vacuum and ~4 for some non-silica glasses (Price et al., 2007) and semiconductors, whereas the nonlinear index n_2 can span many orders of magnitude. A common method for guiding extremely high power pulses is thus in hollow-core capillaries or fibres, where most of the light propagates in a gas. For example, $n_2 \approx 5 \times 10^{-23}$ m^2/W in air, compared to 2.5×10^{-20} m^2/W in silica glass, thus pushing P_{crit} into the GW regime. In contrast to solid-core fibres, gas-filled capillaries do not support strictly bound modes, but all modes are intrinsically leaky with losses scaling proportional to λ^2/R^3 where λ is the light wavelength and R is the radius of the capillary hole (Marcatili & Schmeltzer, 1964). Hence, the capillary hole must be sufficiently large in order to allow for transmission of light over long distances. For example, 800nm wavelength light propagating in the fundamental LP_{01} mode of a silica glass capillary with a 75μm radius hole experiences losses of ~3dB/m. For such a large hole compared to the laser wavelength, the capillary is multimoded, and this is the situation we will consider in the following. It should be noted, however, that single-mode guidance in hollow-core fibres is in principle possible using bandgap effects in photonic crystal fibres (Knight et al., 1998; Petrovich et al., 2008).

Using fs pulses at 800nm wavelength from commercial Ti:sapphire laser systems it is possible
to reach peak powers large enough to observe nonlinear effects, and even self-focusing,
in gases. Capillary guidance is used in this context for several high-power applications.
One of these is pulse compression, where the nonlinearity of the gas in the capillary is
exploited to spectrally broaden a pulse by self-phase modulation, which allows the pulse to
be compressed after the capillary by purely dispersive means such as gratings or dispersive
mirrors (Sartania et al., 1997). For intensities above $\sim 10^{13}$W/cm^2, the electric field of the
laser is large enough to start ionising the gaseous medium. The generated plasma exhibits
a negative refractive index, which can counteract the self-focusing effect of the neutral gas
and lead to pulse filamentation (Couairon & Mysyrowicz, 2007). In another application,
ionisation and recombination effects are used for high harmonic generation of XUV and soft
X-ray radiation, processes whose efficiencies can be enhanced significantly by phase matching
techniques in capillaries (Rundquist et al., 1998).

In the following we will therefore discuss how the MM-NLSE can be extended to include
these important effects and demonstrate a few sample effects related to the multimode nature
of hollow capillaries typically used for such high-power applications.

5.1 Ionisation and plasma effects in the multimode nonlinear Schrödinger equation

The starting point for this derivation is the capability of high-intensity light to ionise the gas
inside the capillary. Two effects contribute to the ionisation: (i) direct multiphoton ionisation,
where several photons are absorbed simultaneously to eject one electron from its orbit, and
(ii) tunneling ionisation, where the electric field of the laser is so strong that it deforms the
electric potential of the nucleus and allows an electron to tunnel through the potential barrier.
Tunneling ionisation occurs at higher field strengths than multiphoton ionisation, and is the
dominant process for the effects we want to discuss here. The rate of tunneling ionisation W
can be calculated using Keldysh theory (Popov, 2004) as

$$W(\mathbf{x},t) = W_0 \kappa^2 \sqrt{\frac{3}{\pi}} C_{\kappa l}^2 2^{2n^*} F(\mathbf{x},t)^{1.5-2n^*} \exp\left(-\frac{2}{3F(\mathbf{x},t)}\right), \tag{12}$$

where $\kappa^2 = I_p/I_H$ is the ratio of the ionisation potential I_p of the gas species over the ionisation
potential for hydrogen $I_H = 13.6$eV, $W_0 = m_e e^4/\hbar^3 = 4.13 \times 10^{16}$s^{-1}, $F(\mathbf{x},t) = E'(\mathbf{x},t)/(\kappa^3 E_a)$
is the reduced electric field of the laser with $E_a = 5.14 \times 10^{11}$V/m the atomic unit of field
intensity and $E'(\mathbf{x},t)$ the real-valued electric field in units of V/m corresponding to $\mathbf{E}(\mathbf{x},t)$,
Eq. (1). The dimensionless parameters $C_{\kappa l}$ and n^* are specific for the gas and can be looked
up in tables (Popov, 2004). For the case of argon, which we will use as our example here, we
have $I_p = 15.76$eV, $C_{\kappa l} = 0.95$, and $n^* = 0.929$.

Given the modal amplitudes $A_p(z,t)$ we can calculate the electric field $\mathbf{E}(\mathbf{x},t)$ and thus the
ionisation rate $W(\mathbf{x},t)$ at every point and time in the capillary. From this we obtain the fraction
of neutral atoms $r_0(\mathbf{x},t)$ and the fraction of ionised atoms $r_1(\mathbf{x},t) = 1 - r_0(\mathbf{x},t)$ by solving

$$\frac{\partial r_0(\mathbf{x},t)}{\partial t} = -W(\mathbf{x},t)r_0(\mathbf{x},t). \tag{13}$$

The generated plasma modifies the refractive index of the gas to

$$n(\mathbf{x},t) = \sqrt{1 - \frac{\omega_{pl}(\mathbf{x},t)^2}{\omega^2}}, \tag{14}$$

where the plasma frequency is given by

$$\omega_{pl}(\mathbf{x}, t) = \sqrt{\frac{\rho r_1(\mathbf{x}, t)e^2}{m_e \epsilon_0}}. \tag{15}$$

Here ρ is the gas density and e and m_e are the electron charge and mass, respectively. The MM-NLSE thus aquires a new nonlinear term $\partial A_p(z,t)/\partial z \propto \mathcal{N}_{pl}\{A_p\}$ with

$$\mathcal{N}_{pl}\{A_p\} = -\left(1 - \frac{i}{\omega_0}\frac{\partial}{\partial t}\right)\frac{i}{2}k_0 \int dxdy \frac{F_p(x,y)^* \cdot \mathbf{E}(\mathbf{x}, t)}{\left[\int dx\, dy\, |F_p|^2\right]^{1/2}} \frac{\omega_{pl}(\mathbf{x}, t)^2}{\omega_0^2}, \tag{16}$$

which includes a self-steepening correction term and the projection of the modified laser field onto mode p via a spatial overlap integral.

In addition to the effect of the plasma induced refractive index, we also have to consider the loss of energy from the propagating laser pulse due the ionisation process itself (Courtois et al., 2001). In the modal decomposition, this leads to a nonlinear loss term in the propagation of the mode envelope A_p of the form

$$\mathcal{L}_{ion}\{A_p\} = -\frac{1}{2}\int dxdy \frac{F_p(x,y)^* \cdot \mathbf{E}(\mathbf{x}, t)}{\left[\int dx\, dy\, |F_p|^2\right]^{1/2}} \frac{\rho r_0(\mathbf{x}, t)W(\mathbf{x}, t)I_p}{|\mathbf{E}(\mathbf{x}, t)|^2}. \tag{17}$$

The full MM-NLSE in the presence of gas ionisation by tunneling in the strong-field limit thus becomes (Chapman et al., 2010)

$$\frac{\partial A_p}{\partial z} = \mathcal{D}\{A_p\} + \mathcal{N}\{A_p\} + \mathcal{N}_{pl}\{A_p\} + \mathcal{L}_{ion}\{A_p\} \tag{18}$$

where the individual terms are given by (2), (3), (16) and (17).

5.2 Ultrashort pulse propagation in capillaries

In the following we present simulation results of the extended MM-NLSE (18) for a specific experimental situation (Froud et al., 2009). In particular, we consider a 7cm long capillary with a 75μm radius hole filled with argon at a pressure of 80mbar in the central 3cm of the capillary; the Ar pressure tapers down over 2cm to 0mbar at the input and output. Laser pulses of 40fs length at 780nm wavelength are launched with a Gaussian waist of 40μm centred into the capillary. For the simulations, 20 linearly polarised LP_{0n} modes are considered, as discussed in Sec. 2.2.

Results from two sets of simulations with different launched pulse energies, 0.5mJ and 0.7mJ, respectively, are presented in Fig. 9. The distribution of Ar^+ ions in the capillary is shown in Figs. 9(a) and (b). As expected, ionisation mainly occurs on axis where the laser intensity is maximum. Moreover, because the transverse beam size of the launched laser pulses is not ideally matched to the fundamental mode of the capillary, power is also coupled into the first higher order mode, which leads to mode beating and thus to the periodic ionisation pattern along the capillary length with a periodicity of ~2cm, observed most clearly at lower powers, Fig. 9(a). At higher powers, the nonlinear ionisation processes become much stronger and a spate of additional radial and longitudinal structures are found in the ionisation pattern, Fig. 9(b). In Fig. 9(c) the partial Ar^+ pressures of (a) and (b) are averaged over the transverse cross section of the capillary. The distribution shown in this figure can be easily verified

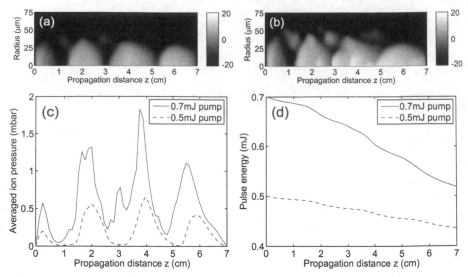

Fig. 9. Propagation of 40fs pulses at 780nm wavelength in a hollow-core capillary (length 7cm, hole radius 75μm) filled with argon with partial ionisation. (a), (b) Partial pressure of Ar$^+$ ions (in dB of mbar) vs position z and radius r inside the capillary for launched pulse energies of 0.5mJ and 0.7mJ, respectively. (c) Ar$^+$ pressure averaged over the capillary cross section vs z. (d) Corresponding integrated pulse energy vs z. The total gas pressure in the capillary centre is 80mbar.

experimentally as it is proportional to the intensity of the Ar$^+$ ion fluorescence observed at 488nm (Chapman et al., 2010; Froud et al., 2009). Finally, in Fig. 9(d) the pulse energy summed over all modes is presented versus the propagation distance for these two simulations. The effect of propagation losses due to ionisation, described by the term $\mathcal{L}_{ion}\{A_p\}$ in Eq. (17), is clearly visible with strong losses associated with the peaks of large ionisation in Fig. 9(c). Because of the highly nonlinear nature of tunneling ionisation, losses at slightly higher input energies (0.7mJ instead of 0.5mJ) are several times larger.

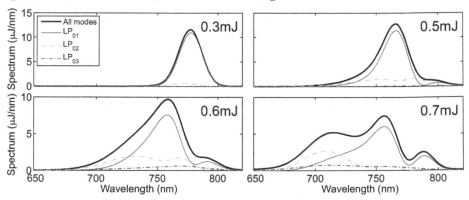

Fig. 10. Pulse spectra and modal contributions at the capillary output for launched pulse energies in the range 0.3mJ to 0.7mJ. Other parameters as in Fig. 9.

The spatial and temporal distribution of ions generated by the propagating laser pulse acts back on the pulse through its (negative) refractive index, according to the term $\mathcal{N}_{pl}\{A_p\}$ given in Eq. (16). Because of the strong localisation of the regions with high ionisation, different capillary modes are affected differently resulting in strong intermodal scattering and mode-specific spectral broadening, as is demonstrated in Fig. 10. At a relatively low pulse energy of 0.3mJ where ionisation is weak, a slight blue-shift of the spectral contribution of the excited LP_{02} mode is observed, but no higher order mode excitation. Increasing the pulse energy to 0.5-0.7mJ, more and more light is scattered into higher order modes. Moreover, the spectrum first develops a small peak at the long-wavelength side of the pump (790-800nm) and then a very broad and high-intensity shoulder at short wavelengths. It is interesting to note that these short wavelength parts of the spectrum are more pronounced in the higher order modes LP_{02} and LP_{03} of the capillary, in fact they contain more power than the fundamental mode at these wavelengths for launched pulse energies above 0.6mJ. This finding has again been confirmed by experiments, where a strong position-dependence of the spectrum was observed in the far field beyond the capillary (Chapman et al., 2010).

These selected results demonstrate clearly that mode interference and mode coupling, i.e., transverse spatial effects, play a significant role in the propagation of high-intensity laser pulses in regimes where ionisation becomes important. This also impacts other applications of such systems, for example the angular dependence of high harmonic generation as recently observed in a capillary-based XUV source (Praeger et al., 2007).

6. Conclusions and outlook

To summarise, we presented an analysis of nonlinear effects of short laser pulses propagating in multimode optical fibres. We developed a general theoretical framework which is based on the modal decomposition of the propagating light and takes the form of a multimode generalised nonlinear Schrödinger equation. This approach provides new insights into the significance of fibre properties, e.g., modal dispersion and mode overlaps, for nonlinear pulse propagation, and for moderately multimode fibres and waveguides it has been shown to be numerically significantly more efficient than beam propagation methods. We subsequently discussed several applications of the model covering laser peak powers in the kW (supercontinuum generation), MW (self-focusing effects) and GW regime (ionisation and plasma nonlinearities) highlighting the importance of multimode effects throughout.

While we focused our discussion here on the high-power regime, we emphasise that there is also rapidly growing interest in the application of multimode fibres at low, W-level peak powers. A fast emerging area of interest comes, for example, from optical telecommunications, where in an attempt to increase the fibre capacity researchers are now considering the use of several fibre modes, or several cores within a single fibre, as independent channels. Intermodal nonlinear effects are expected to pose an ultimate limit to the maximum information capacity of the link, which we believe could be estimated by simulations using our model. Various sensing and imaging applications can also benefit from multimode fibres. Moreover, new sources in the mid-IR spectral region are currently being developed for spectroscopy and sensing applications that require novel waveguides such as soft glass fibres or semiconductor-based waveguides and fibres, some of which are intrinsically multimoded at near-IR pump wavelengths. We therefore expect that the multimode nonlinear Schrödinger equation discussed in this work will provide a valuable tool in the analysis and investigation of many future photonics applications.

7. Acknowledgements

The authors acknowledge financial support by the U.K. Engineering and Physical Sciences Research Council (EPSRC) and the Royal Society. We thank Prof. D. J. Richardson, Dr. W. S. Brocklesby, and Prof. J. G. Frey for valuable discussions.

8. References

Agrawal, G. P. (2001). *Nonlinear Fiber Optics*, 3rd ed., Academic Press, San Diego, USA.

Askaryan, G. A. (1962). Effects of the gradient of strong electromagnetic beam on electrons and atoms, *Sov. Phys. JETP*, Vol. 15, 1088–1090.

Blow, K. J., & Wood, D. (1989). Theoretical description of transient stimulated Raman scattering in optical fibers. *IEEE J. Quantum Electron.*, Vol. 25, No. 12, 2665–2673.

Brabec, T., & Krausz, F. (1997). Nonlinear optical pulse propagation in the single-cycle regime. *Phys. Rev. Lett.*, Vol. 78, No. 17, 3282–3285.

Chaipiboonwong, T., Horak, P., Mills, J. D., & Brocklesby, W. S. (2007). Numerical study of nonlinear interactions in a multimode waveguide. *Opt. Expr.*, Vol. 15, 9040–9047.

Chapman, R. T., Butcher, T. J., Horak, P., Poletti, F., Frey, J. G., & Brocklesby, W. S. (2010). Modal effects on pump-pulse propagation in an Ar-filled capillary. *Opt. Expr.*, Vol. 18, No. 12, 13279–13284.

Chiao, R. Y., Garmire, E., & Townes, C. H. (1964). Self-trapping of optical beams. *Phys. Rev. Lett.*, Vol. 13, No. 15, 479–482.

Coen, S., Chau, A. H. L., Leonhardt, R., Harvey, J. D., Knight, J. C., Wadsworth, W. J., & Russell, P. St. J. (2002). Supercontinuum generation by stimulated Raman scattering and parametric four-wave mixing in photonic crystal fibers. *J. Opt. Soc. Am. B*, Vol. 19, No. 4, 753–764.

Couairon, A., & Mysyrowicz, A. (2007). Femtosecond filamentation in transparent media. *Phys. Rep.*, Vol. 441, 47–189.

Courtois, C., Couairon, A., Cros, B., Marquès, J. R., & Matthieussent, G. (2001). Propagation of intense ultrashort laser pulses in a plasma filled capillary tube: Simulations and experiments. *Phys. Plas.*, Vol. 8, No. 7, 3445–3456.

Delmonte, T., Watson, M. A., O'Driscoll, E. J., Feng, X., Monro, T. M., Finazzi, V., Petropoulos, P., Price, J. H. V., Baggett, J. C., Loh, W., Richardson, D. J., & Hand, D. P. (2006). Generation of mid-IR continuum using tellurite microstructured fiber. *Conference on Lasers and Electro-Optics*, paper CTuA4, Long Beach, USA.

Dudley, J. M., Provino, L., Grossard, N., Maillotte, H., Windeler, R. S., Eggleton, B. J., & Coen, S. (2002). Supercontinuum generation in air-silica microstructured fibers with nanosecond and femtosecond pulse pumping. *J. Opt. Soc. Am. B*, Vol. 19, No. 4, 765–771.

Dudley, J. M., Genty, G., & Coen, S. (2006). Supercontinuum generation in photonic crystal fiber. *Rev. Mod. Phys.*, Vol. 78, No. 4, 1135–1184.

Dupriez, P., Poletti, F., Horak, P., Petrovich, M. N., Jeong, Y., Nilsson, J., Richardson, D. J., & Payne, D. N. (2007). Efficient white light generation in secondary cores of holey fibers. *Opt. Expr.*, Vol. 15, No. 7, 3729–3736.

Farrow, R. L., Kliner, D. A. V., Hadley, G. R., & Smith, A. V. (2006). Peak-power limits on fiber amplifiers imposed by self-focusing. *Opt. Lett.*, Vol. 31, No. 23, 3423–3425.

Fibich, G., & Gaeta, A. (2000). Critical power for self-focusing in bulk media and in hollow waveguides. *Opt. Lett.*, Vol. 25, No. 5, 335–337.

Froud, C. A., Chapman, R. T., Rogers, E. T. F., Praeger, M., Mills, B., Grant-Jacob, J., Butcher, T. J. , Stebbings, S. L., de Paula, A. M., Frey, J. G., & Brocklesby, W. S. (2009). Spatially resolved Ar* and Ar^{+*} imaging as a diagnostic for capillary-based high harmonic generation. *J. Opt. A*, Vol. 11, 054011.

Gaeta, A. (2000). Catastrophic collapse of ultrashort pulses. *Phys. Rev. Lett.*, Vol. 84, No. 16, 3582–3585.

Galvanauskas, A., Cheng, M.-Y., Hou, K.-C., Liao, K.-H. (2007). High peak power pulse amplification in large-core Yb-doped fiber amplifiers. *IEEE J. Sel. Top. Quantum Electron.*, Vol. 13, No. 3, 559–566.

Hasegawa, A., & Tappert, F. (1973). Transmission of stationary nonlinear optical pulses in dispersive dielectric fibers. I. Anomalous dispersion. *Appl. Phys. Lett.*, Vol. 23, No. 3, 142–144.

Horak, P., & Poletti, F. (2009). Effects of pulse self-focusing on supercontinuum generation in multimode optical fibers. *International Conference on Transparent Optical Networks*, Ponta Delgada, Portugal, 28 June - 2 July 2009.

Jeong, Y., Sahu, J. K., Payne, D. N., & Nilsson, J. (2004). Ytterbium-doped large-core fiber laser with 1.36 kW continuous-wave output power. *Opt. Expr.*, Vol. 12, No. 25, 6088–6092.

Kao, K. C., & Hockham, G. A. (1966). Dielectric-fibre surface waveguides for optical frequencies. *Proc. IEE*, Vol. 113, No. 7, 1151–1158.

Knight, J. C., Broeng, J., Birks, T. A., & Russell, P. St. J. (1998). Photonic band gap guidance in optical fibers. *Science*, Vol. 282, 1476–1478.

Lehtonen, M., Genty, G., Ludvigsen, H., & Kaivola, M. (2003). Supercontinuum generation in a highly birefringent microstructured fiber. *Appl. Phys. Lett.*, Vol. 82, No. 14, 2197–2199.

Lesvigne, C., Couderc, V., Tonello, A., Leproux, P., Barthelemy, A., Lacroix, S., Druon, F., Blandin, P., Hanna, M., & Georges, P. (2007). Visible supercontinuum generation controlled by intermodal four-wave mixing in microstructured fiber. *Opt. Lett.*, Vol. 32, No. 15, 2173–2175.

Marcatili, E. A. J., & Schmeltzer, R. A. (1964). Hollow metallic and dielectric waveguides for long distance optical transmission and lasers. *Bell Tech. Syst. J.*, Vol. 43, 1783–1809.

Martins, E. R., Spadoti, D. H., Romero, M. A., & Borges, B.-H. V. (2007). Theoretical analysis of supercontinuum generation in a highly birefringent D-shaped microstructured optical fiber. *Opt. Expr.*, Vol. 15, No. 22, 14335–14347.

Mears, R. J., Reekie, L., Jauncey, I. M., & Payne, D. N. (1987). Low-noise Erbium-doped fiber amplifier at 1.54μm. *Electron. Lett.*, Vol. 23, No. 19, 1026–1028.

Milosevic, N., Tempea, G., & Brabec, T. (2000). Optical pulse compression: bulk media versus hollow waveguides. *Opt. Lett.*, Vol. 25, No. 9, 672–674.

Miya, T., Terunuma, Y., Hosaka, T., & Miyashita, T. (1979). Ultimate low-loss single-mode fibre at 1.55μm. *Electron. Lett.*, Vol. 15, No. 4, 106–108.

Mortensen, N. A., Nielsen, M. D., Folkenberg, J. R., Petersson, A., & Simonsen, H. R. (2003). Improved large-mode-area endlessly single-mode photonic crystal fibers. *Opt. Lett.*, Vol. 28, No. 6, 393–395.

Okamoto, K. (2006). *Fundamentals of Optical Waveguides*, 2nd ed., Academic Press, San Diego, USA.

Petrovich, M. N., Poletti, F., van Brakel, A., & Richardson, D. J. (2008). Robustly single mode hollow core photonic bandgap fiber. *Opt. Expr.*, Vol. 16, No. 6, 4337–4346.

Poletti, F., & Horak, P. (2008). Description of ultrashort pulse propagation in multimode optical fibers. *J. Opt. Soc. Am. B*, Vol. 25, No. 10, 1645–1654.

Poletti, F., & Horak, P. (2009). Dynamics of femtosecond supercontinuum generation in multimode fibers. *Opt. Expr.*, Vol. 17, No. 8, 6134–6147.

Popov, V. S. (2004). Tunnel and multiphoton ionization of atoms and ions in a strong laser field (Keldysh theory). *Physics-Uspekhi*, Vol. 47, No. 9, 855–885.

Praeger, M., de Paula, A. M., Froud, C. A., Rogers, E. T. F., Stebbings, S. L., Brocklesby, W. S., Baumberg, J. J., Hanna, D. C., & Frey, J. G. (2007). Spatially resolved soft X-ray spectrometry from single-image diffraction. *Nat. Phys.*, Vol. 3, 176–179.

Press, W. H., Teukolsky, S. A., Vetterling, W. T., & Flannery, B. P. *Numerical recipes: the art of scientific computing*, 3rd ed. (Cambridge University Press, New York, USA, 2007).

Price, J. H. V., Monro, T. M., Furusawa, K., Belardi, W., Baggett, J. C., Coyle, S., Netti, C., Baumberg, J. J., Paschotta, R., & Richardson, D. J. (2003). UV generation in a pure-silica holey fiber. *Appl. Phys. B*, Vol. 77, No. 2–3, 291–298.

Price, J. H. V., Monro, T. M., Ebendorff-Heidepriem, H., Poletti, F., Horak, P., Finazzi, V., Leong, J. Y. Y., Petropoulos, P., Flanagan, J. C., Brambilla, G., Feng, X., & Richardson, D. J. (2007). Mid-IR supercontinuum generation from nonsilica microstructured optical fibers. *J. Sel. Top. Quantum Electron.*, Vol. 13, No. 3, 738–749.

Ranka, J. K., & Gaeta, A. L. (2009). Breakdown of the slowly varying envelope approximation in the self-focusing of ultrashort pulses. *Opt. Lett.*, Vol. 23, No. 7, 534–536.

Roppo, V., Centini, M., Sibilia, C., Bertolotti, M., de Ceglia, D., Scalora, M., Akozbek, N., Bloemer, M. J., Haus, J. W., Kosareva, O. G., & Kandidov, V. P. (2007). Role of phase matching in pulsed second-harmonic generation: Walk-off and phase-locked twin pulses in negative-index media. *Phys. Rev. A*, Vol. 76, No. 3, 033829.

Rundquist, A., Durfee, C. G., Chang, Z., Herne, C., Backus, S., Murnane, M. M., & Kapteyn, H. C. (1998). Phase-matched generation of coherent soft X-rays. *Science*, Vol. 280, 1412–1415.

Sartania, S., Cheng, Z., Lenzner, M., Tempea, G., Spielmann, Ch., Krausz, F., and Ferencz, K. (1997). Generation of 0.1-TW 5-fs optical pulses at a 1-kHz repetition rate. *Opt. Lett.*, Vol. 22, No. 20, 1562–1564.

Schreiber, T., Andersen, T., Schimpf, D., Limpert, J., & Tünnermann, A. (2005). Supercontinuum generation by femtosecond single and dual wavelength pumping in photonic crystal fibers with two zero dispersion wavelengths. *Opt. Expr.*, Vol. 13, No. 23, 9556–9569.

Stolen, R. H., Ippen, E. P., & Tynes, A. R. (1972). Raman oscillation in glass optical waveguide. *Appl. Phys. Lett.*, Vol. 20, No. 2, 62–64.

Stolen, R. H., & Bjorkholm, J. B. (1982). Parametric amplification and frequency conversion in optical fibers. *IEEE J. Quantum Electron.*, Vol. 18, No. 7, 1062–1072.

Stuart, B. C., Feit, M. D., Herman, S., Rubenchik, A. M., Shore, B. W., & Perry, M. D. (1996). Nanosecond-to-femtosectond laser-induced breakdown in dielectrics. *Phys. Rev. B*, Vol. 53, No. 4, 1749–1761.

Tonello, A., Pitois, S., Wabnitz, S., Millot, G., Martynkien, T., Urbanczyk, W., Wojcik, J., Locatelli, A., Conforti, M., & De Angelis, C. (2006). Frequency tunable polarization and intermodal modulation instability in high birefringence holey fiber. *Opt. Expr.*, Vol. 14, No. 1, 397–404.

Zharova, N. A., Litvak, A. G., & Mironov, V. A. (2006). Self-focusing of wave packets and envelope shock formation in nonlinear dispersive media. *J. Exp. Theor. Phys.*, Vol. 103, No. 1, 15–22.

2

Optical Solitons in a Nonlinear Fiber Medium with Higher-Order Effects

Deng-Shan Wang

School of Science, Beijing Information Science and Technology University, Beijing, 100192 China

1. Introduction

Nowadays we can see many interesting applications of solitons in different areas of physical sciences such as plasma physics (1), nonlinear optics (2; 3), Bose-Einstein condensate (4; 5), fluid mechanics (6), and so on. Solitons are so robust particles that they are unlikely to breakdown under small perturbations. The most interesting factor about the soliton, however, is that their interactions with the medium through which it propagates is elastic. Recent researches on nonlinear optics have shown that dispersion-managed pulse can be more useful if the pulse is in the form of a power series of a stable localized pulse which is called soliton.

Optical solitons have been the objects of extensive theoretical and experimental studies during the last four decades, because of their potential applications in long distance communication. In 1973, the pioneering results of Hasegawa and Tappert (7) proved that the major constraint in the optical fiber, namely, the group velocity dispersion (GVD) could be exactly counterbalanced by the self-phase modulation (SPM). SPM is the dominant nonlinear effect in silica fibers due to the Kerr effect. The theoretical results of Hasegawa and Tappert were greatly supported by the experimental demonstration of optical solitons by Mollenauer et al. (8) in 1980. Since then many theoretical and experimental works have been done to achieve a communication system based on optical solitons.

The solitons, localized-in-time optical pulses, evolve from a nonlinear change in the refractive index of the material, known as Kerr effect, induced by the light intensity distribution. When the combined effects of the intensity-dependent refractive index nonlinearity and the frequency-dependent pulse dispersion exactly compensate for one another, the pulse propagates without any change in its shape, being self-trapped by the waveguide nonlinearity. The propagation of optical solitons in a nonlinear dispersive optical fiber is governed by the well-known completely integrable nonlinear Schrödinger (NLS) equation

$$i\frac{\partial q}{\partial z} + \epsilon\frac{\partial^2 q}{\partial \tau^2} + |q|^2 q = 0, \qquad \epsilon = \pm 1, \tag{1}$$

where q is the complex amplitude of the pulse envelope, τ and z represent the spatial and temporal coordinates, and the $+$ and $-$ sign of ϵ before the dispersive term denote the anomalous and normal dispersive regimes, respectively. In the anomalous dispersive regime, this equation possesses a bright soliton solution, and in the normal dispersive regime it possesses dark solitons. The bright soliton and dark soliton solutions can be derived by the inverse-scattering transform method with vanishing (9; 11) and nonvanishing boundary conditions (10).

However, if optical pulses are shorter, the standard NLS equation becomes inadequate. Therefore, some higher-order effects such as third-order dispersion, self-steepening, and stimulated Raman scattering, will play important roles in the propagation of optical pulses. In such a case, the governing equation is the one known widely as the higher-order NLS equation, first derived by Kodama and Hasegawa (12). The effect of these effects in uncoupled and coupled systems for bright solitons is well explained (13; 14). Inelastic Raman scattering is due to the delayed response of the medium, which forces the pulse to undergo a frequency shift which is known as a self-frequency shift. The effect of self-steepening is due to the intensity-dependent group velocity of the optical pulse, which gives the pulse a very narrow width in the course of propagation. Because of this, the peak of the pulse will travel more slowly than the wings.

In practice, the refractive index or the core diameter of the optical fiber are fucntions of the axial coordinate, which means that the fiber is actually axially inhomogeneous. In this case, the parameters which characterize the dispersive and nonlinear properties of the fiber exhibit variations and the corresponding nonlinear wave equations are NLS equations with variable coefficients. Moreover, the problem of ultrashort pulse propagation in nonlinear and axially inhomogeneous optical fibers near the zero dispersion point is more complicated because the high order effects have to be taken into account as well. In order to understand such phenomena, we consider the higher-order NLS (HNLS) equation with variable coefficients

$$\frac{\partial u}{\partial z} = i(d_1 \frac{\partial^2 u}{\partial \tau^2} + d_2 |u|^2 u) + d_3 \frac{\partial^3 u}{\partial \tau^3} + d_4 \frac{\partial(u|u|^2)}{\partial \tau} + d_5 u \frac{\partial |u|^2}{\partial \tau} + d_6 u, \tag{2}$$

where u is the slowly varying envelope of the pulse, d_1, d_2, d_3, d_4, d_5 and d_6 are the z-dependent real parameters related to GVD, SPM, third-order dispersion (TOD), self-steepening, and stimulated Raman scattering (SRS), and the heat-insulating amplification or loss, respectively. Though Eq. (2) was first derived in the year 1980s, only for the past few years, it has attracted much attention among the researchers from both theoretical and experimental points of view. For example, Porsezian and Nakkeeran (13) derive all parametric conditions for soliton-type pulse propagation in HNLS equation using the Painlevé analysis, and generalize the Ablowitz-Kaup-Newell-Segur method to the 3×3 eigenvalue problem to construct the Lax pair for the integrable case. Papaioannou et al. (15) give an analytical treatment of the effect of axial inhomogeneity on femtosecond solitary waves near the zero dispersion point which governed by the variable-coefficient HNLS equation. The exact bright and dark soliton wave solutions of this variable-coefficient equation are derived and their behaviors in the presence of the inhomogeneity are analyzed. Mahalingam and Porsezian (16) analyze the propagation of dark solitons with higher-order effects in optical fibers by Painlevé analysis and Hirota bilinear method. Xu et al. (17) investigate the modulation instability and solitons on a cw background in an optical fiber with higher-order effects. In addition, there have recently been several papers giving W-shaped solitary wave solution in the HNLS equation. However, in recent years the studies of Eq. (2) have not been widespread. In this chapter, we consider equation (2) again and derive some exact soliton solutions in explicit form for specified soliton management conditions. We first change the variable-coefficient HNLS equation into the well-known constant-coefficient HNLS equation through similarity transformation. Then the Lax pairs for two integrable cases of the constant-coefficient HNLS equation are constructed explicitly by prolongation technique, and the novel exact bright N-soliton solutions for the bright soliton version of HNLS equation are obtained by Riemann-Hilbert formulation. Finally, we examine the dynamics and present the features of the optical solitons. It is seen that the bright two-soliton solution of the HNLS equation behaves in an elastic manner characteristic of all soliton solutions. These results are useful in the design of transmission lines with spatial parameter variations and soliton management to future research.

2. Similarity transformation

A direct and efficient method for investigating the variable-coefficient nonlinear wave equation is to transform them into their constant-coefficient counterparts by similarity transformation. To do so, we firstly take the similarity transformation (18; 19)

$$u = \rho q (T, X) e^{i(a_1 \tau + a_2)},$$ (3)

to reduce Eq. (2) to the constant-coefficient HNLS equation

$$\frac{\partial q}{\partial T} = i(\alpha_1 \frac{\partial^2 q}{\partial X^2} + \alpha_2 |q|^2 q) + \epsilon(\alpha_3 \frac{\partial^3 q}{\partial X^3} + \alpha_4 \frac{\partial(q|q|^2)}{\partial X} + \alpha_5 q \frac{\partial |q|^2}{\partial X}),$$ (4)

where $q = q(T, X)$ is the complex amplitude of the pulse envelope, the parameter ϵ ($0 < \epsilon < 1$) denotes the relative width of the spectrum that arises due to the quasi-monochromocity, $\alpha_1, \alpha_2, \alpha_3, \alpha_4$ and α_5 are the real constant parameters. In Eq. (3), ρ, T, a_1 and a_2 are functions of z, and X is a function of τ and z.

Substituting Eq. (3) into Eq. (2) and asking $q (T, X)$ to satisfy the constant-coefficient HNLS equation (4), we have a set of partial differential equations (PDEs)

$$d_1 X_{\tau\tau} + 3 d_3 X_{\tau\tau} a_1 = 0, \quad d_3 X_\tau^3 = \alpha_3 T_z, \quad \rho_z = d_6 \rho,$$

$$2 d_1 X_\tau a_1 + X_z + 3 d_3 X_\tau a_1^2 = d_3 X_{\tau\tau\tau}, \quad \rho^2 d_4 a_1 + \rho^2 d_2 = \alpha_2 T_z,$$
$$2\rho^2 X_\tau d_4 + \rho^2 X_\tau d_5 = 2 \alpha_4 T_z + \alpha_5 T_z, \quad \rho^2 X_\tau d_4 + \rho^2 X_\tau d_5 = \alpha_4 T_z + \alpha_5 T_z,$$
$$d_3 a_1^3 + a_{1z}\tau + a_{2z} + d_1 a_1^2 = 0, \quad 3 d_3 X_\tau^2 a_1 + d_1 X_\tau^2 = \alpha_1 T_z, \quad X_{\tau,\tau} = 0,$$

where the subscript denotes the derivative with respect to z and τ. Solving this set of PDEs, we have $X = k\tau + f$ and

$$a_1 = c, \quad d_1 = \frac{T_z (k\alpha_1 - 3 \alpha_3 c)}{k^3}, \quad d_2 = \frac{T_z (\alpha_2 k - \alpha_4 c)}{\rho^2 k}, \quad d_3 = \frac{\alpha_3 T_z}{k^3},$$

$$d_4 = \frac{\alpha_4 T_z}{\rho^2 k}, \quad d_5 = \frac{\alpha_5 T_z}{\rho^2 k}, \quad f = \frac{c (3 \alpha_3 c - 2 k\alpha_1) T}{k^2}, \quad a_2 = \frac{(2 \alpha_3 c - k\alpha_1) c^2 T}{k^3},$$

where $\rho = \rho_0 e^{\int d_6 z}$, k, ρ_0 and c are constants, and T and d_6 are arbitrary functions of z. So the similarity transformation (3) becomes

$$u = \rho_0 e^{\int d_6 dz} q \left(T, \frac{k^3 \tau - 2 c k\alpha_1 T + 3 c^2 \alpha_3 T}{k^2} \right) e^{ic(k^3\tau + 2 c^2\alpha_3 T - ck\alpha_1 T)/k^3}.$$ (5)

Therefore, if we can get the exact soliton solutions of the constant-coefficient HNLS equation (4) we can obtain the exact soliton solutions for HNLS equation (2) through Eq. (5). In the next section, we will investigate the integrable condition of equation (4) by prolongation technique.

3. Prolongation structures of the constant-coefficient HNLS equation

In this section, we investigate the prolongation structures of the constant-coefficient HNLS equation (4) by means of the prolongation technique (20–22). Firstly, the complex conjugate of the dependent variable q in Eq. (4) is denoted as $q^* = u$. Then, Eq. (4) and its conjugate become

$$i\alpha_1 q_{XX} + i\alpha_2 q^2 u + \epsilon[\alpha_3 q_{XXX} + (\alpha_4 + \alpha_5)q^2 u_X + (2\alpha_4 + \alpha_5)quq_X] - q_T = 0, \tag{6a}$$

$$-i\alpha_1 u_{XX} - i\alpha_2 u^2 q + \epsilon[\alpha_3 u_{XXX} + (\alpha_4 + \alpha_5)u^2 q_X + (2\alpha_4 + \alpha_5)quu_X] - u_T = 0. \tag{6b}$$

Next we introduce four new variables p, r, v and w by

$$q_X = p, \quad p_X = r, \quad u_X = v, \quad v_X = w, \tag{7}$$

and define a set of differential 2-form $I = \{\theta_1, \theta_2, \theta_3, \theta_4, \theta_5, \theta_6\}$ on solution manifold $M = \{T, X, u, v, w, p, q, r\}$, where

$$\theta_1 = dq \wedge dT + pdT \wedge dX, \quad \theta_2 = dp \wedge dT + rdT \wedge dX,$$

$$\theta_3 = du \wedge dT + vdT \wedge dX, \quad \theta_4 = dv \wedge dT + wdT \wedge dX,$$

$$\theta_5 = dq \wedge dX + \alpha_3 dr \wedge dT + \rho_1 dX \wedge dT, \quad \theta_6 = du \wedge dX + \alpha_3 dw \wedge dT + \rho_2 dX \wedge dT,$$

with

$$\rho_1 = i\alpha_1 r + i\alpha_2 q^2 u + \epsilon[(\alpha_4 + \alpha_5)q^2 v + (2\alpha_4 + \alpha_5)qup],$$

$$\rho_2 = -i\alpha_1 w - i\alpha_2 u^2 q + \epsilon[(\alpha_4 + \alpha_5)u^2 p + (2\alpha_4 + \alpha_5)quv].$$

When these differential 2-forms restricted on the solution manifold M become zero, we recover the original constant-coefficient HNLS equation (4). It is easy to verify that I is a differential closed idea, i.e. $dI \subset I$.

We further introduce n differential 1-forms

$$\Omega^i = d\zeta^i - \tilde{F}^i dX - \tilde{G}^i dT, \tag{8}$$

where $i = 1, 2, \cdots, n$, $\tilde{F}^i$ and $\tilde{G}^i$ are functions of $u, v, w, p, q, r, \zeta^i$ and are assumed to be both linearly dependent on ζ^i, namely $\tilde{F}^i = F^i \zeta^i, \tilde{G}^i = G^i \zeta^i$. For the sake of simplification, we drop the indices by rewriting ζ^i as ζ, F^i as F and G^i as G. When restricting on solution manifold, the differential 1-forms Ω^i are null, i.e. $\Omega^i = 0$ which is just the linear spectral problem $\zeta_X = F\zeta$ and $\zeta_T = G\zeta$.

Following the well-known prolongation technique, the extended set of differential form $\tilde{I} = I \cup \{\Omega^i\}$ must be a closed ideal under exterior differentiation, i.e. $d\tilde{I} \subset \tilde{I}$. Because $dI \subset I \subset \tilde{I}$, we only need to let $d\{\Omega^i\} \subset \tilde{I}$, which denotes that

$$d\Omega^i = \sum_{j=1}^{6} f_j^i \theta^j + \eta^i \wedge \Omega^i, \quad i = 1, 2, \cdots, n, \tag{9}$$

where f_j^i ($j = 1, 2, 3, 4, 5, 6$) are functions of (T, X), and $\eta^i = g^i(T, X)dX + h^i(T, X)dT$ are differential 1-forms.

When Eq. (9) is written out in detail, after dropping the indices we have the following PDEs about F and G as

$$G_r = \epsilon \alpha_3 F_q, \quad G_w = \epsilon \alpha_3 F_u,$$

$$G_q p + G_p r + G_u v + G_v w - F_q \left[i\alpha_1 r + i\alpha_2 q^2 u + \epsilon (\alpha_4 + \alpha_5) q^2 v + \epsilon (2\alpha_4 + \alpha_5) q u p \right] \quad (10)$$

$$-F_u \left[-i\alpha_1 w - i\alpha_2 u^2 q + \epsilon (\alpha_4 + \alpha_5) u^2 p + \epsilon (2\alpha_4 + \alpha_5) q u v \right] - [F, G] = 0,$$

with $[F, G] = FG - GF$.
Solving Eq. (10), we have the expressions of F and G as

$$F = x_0 + x_1 q + x_2 u, \quad (11)$$

$$G = \epsilon \alpha_3 x_1 r + \epsilon \alpha_3 x_2 w + v q \epsilon \alpha_3 x_5 + v \epsilon \alpha_3 x_4 - p u \epsilon \alpha_3 x_5 + p \epsilon \alpha_3 x_3 - i q u \alpha_1 x_5$$

$$+ i p x_1 \alpha_1 + q x_2 \epsilon \, u^2 \alpha_4 + \frac{2}{3} q x_2 \epsilon \, u^2 \alpha_5 + \frac{1}{2} \epsilon \alpha_3 u^2 x_{13} + q^2 \epsilon \, x_1 u \alpha_4 + \frac{2}{3} q^2 \epsilon \, x_1 u \alpha_5$$

$$+ q u \epsilon \, \alpha_3 x_8 + q u \epsilon \, \alpha_3 x_{10} + \epsilon \, \alpha_3 x_7 u + \frac{1}{2} q^2 \epsilon \, \alpha_3 x_9 + q \epsilon \, \alpha_3 x_6 + i q \alpha_1 x_3 - i \alpha_1 x_4 u - i v x_2 \alpha_1 + x_{15},$$

where $L = \{x_0, x_1, x_2, \cdots, x_{15}\}$ is an incomplete Lie algebra which is called prolongation algebra and it satisfies the following commutation relations

$$[x_2, x_5] = x_{14}, \quad x_2 \alpha_5 = 3 \alpha_3 x_{14}, 2 x_8 + x_{10} = x_{12}, x_1 \alpha_5 + 3 \alpha_3 x_{11} = 0,$$

$$[x_0, x_1] = x_3, \quad [x_0, x_2] = x_4, \quad [x_0, x_3] = x_6, \quad [x_0, x_4] = x_7, \quad [x_0, x_5] = x_8,$$

$$[x_1, x_2] = x_5, \quad [x_1, x_3] = x_9, \quad [x_1, x_4] = x_{10}, \quad [x_1, x_5] = x_{11}, \quad [x_2, x_3] = x_{12},$$

$$[x_2, x_4] = x_{13}, \quad \epsilon \alpha_3 [x_0, x_9] + 2 i\alpha_1 x_9 + 2 \epsilon \alpha_3 [x_1, x_6] = 0, \quad [x_1, x_{15}] + i\alpha_1 x_6 + \epsilon \alpha_3 [x_0, x_6],$$

$$\alpha_3 [x_1, x_9] = 0, \quad \alpha_3 [x_2, x_{13}] = 0, \quad [x_0, x_{15}] = 0, \quad 2 \epsilon \alpha_3 [x_2, x_7] + \epsilon \alpha_3 [x_0, x_{13}] = 2 i\alpha_1 x_{13},$$

$$[x_2, x_{15}] + \epsilon \alpha_3 [x_0, x_7] = i\alpha_1 x_7, \quad 6 \epsilon \alpha_3 ([x_0, x_{10}] + [x_1, x_7] + [x_2, x_6] + [x_0, x_8]) + 6 i\alpha_1 x_8 = 0,$$

$$(6 \epsilon [x_1, x_8] + 6 \epsilon [x_1, x_{10}] + 3 \epsilon [x_2, x_9]) \alpha_3^2 + (4 \epsilon \alpha_5 x_3 + 6 \epsilon \alpha_4 x_3 + 6 i x_1 \alpha_2) \alpha_3 + 2 i\alpha_1 x_1 \alpha_5 = 0,$$

$$(6 \epsilon [x_2, x_{10}] + 3 \epsilon [x_1, x_{13}] + 6 \epsilon [x_2, x_8]) \alpha_3^2 + (4 \epsilon \alpha_5 x_4 + 6 \epsilon \alpha_4 x_4 - 6 i x_2 \alpha_2) \alpha_3 - 2 i\alpha_1 x_2 \alpha_5 = 0.$$

It is known that nontrivial matrix representations of prolongation algebra L correspond to nontrivial prolongation structures. To find the matrix representation of L, following the procedure of Fordy (23), we try to embed it into Lie algebra $sl(n, C)$. Starting from the case of $n = 2$, we found that $sl(2, C)$ is the whole algebra for some special coefficients $\alpha_j (j = 1, 2, 3, 4, 5)$. For the case of $n = 3$, we can also find that $sl(3, C)$ will be the whole algebra for some other special coefficients $\alpha_j (j = 1, 2, 3, 4, 5)$. In this paper, we only examine the case of $sl(2, C)$ algebra.
From the above commutation relations, we have the special relations among elements x_1, x_2 and x_5 as

$$[x_2, x_5] = \frac{\alpha_5}{3\alpha_3} x_2, \quad [x_1, x_5] = -\frac{\alpha_5}{3\alpha_3} x_1, \quad [x_1, x_2] = x_5, \quad (12)$$

from which we know that x_1 and x_2 are nilpotent elements and x_5 is a neutral element. So we have $\alpha_5 = \pm 6 \delta^2 \alpha_3$ and

$$x_1 = \begin{pmatrix} 0 & \delta \\ 0 & 0 \end{pmatrix}, \quad x_2 = \begin{pmatrix} 0 & 0 \\ \pm\delta & 0 \end{pmatrix}, \quad x_5 = \begin{pmatrix} \pm\delta^2 & 0 \\ 0 & \mp\delta^2 \end{pmatrix}, \quad (13)$$

with δ a nonzero constant. Substituting (13) into the commutation relations of prolongation algebra L, we finally get the 2×2 matrix representations of F and G. Therefore, we obtain two integrable HNLS equations with 2×2 spectral problems.

When $\alpha_2 = 2\delta^2\alpha_1, \alpha_4 = 6\delta^2\alpha_3$ and $\alpha_5 = -6\delta^2\alpha_3$, Eq. (4) becomes the bright soliton version of Hirota equation

$$q_T = i\alpha_1 q_{XX} + 2i\alpha_1\delta^2|q|^2 q + \epsilon\,\alpha_3 q_{XXX} + 6\epsilon\delta^2\alpha_3\,|q|^2 q_X, \tag{14}$$

with linear spectral problem

$$\zeta_X = F\zeta, \quad \zeta_T = G\zeta, \tag{15}$$

and

$$F = \begin{pmatrix} -i\lambda & \delta q \\ -\delta q^* & i\lambda \end{pmatrix}, \tag{16}$$

$$G = 4i\alpha_3\epsilon\lambda^3 \begin{pmatrix} 1 & 0 \\ 0 & -1 \end{pmatrix} - 2\lambda^2 \begin{pmatrix} i\alpha_1 & 2\epsilon\,\alpha_3\delta\,q \\ -2\epsilon\,\alpha_3\delta q^* & -i\alpha_1 \end{pmatrix}$$

$$+ 2\lambda \begin{pmatrix} -i\epsilon\,\alpha_3\delta^2|q|^2 & \alpha_1\delta q - i\epsilon\,\alpha_3\delta q_X \\ -\alpha_1\delta q^* - i\epsilon\,\alpha_3\delta q_X^* & i\epsilon\,\alpha_3\delta^2|q|^2 \end{pmatrix} \tag{17}$$

$$+ \begin{pmatrix} \epsilon\,\alpha_3\delta^2 q_X q^* - \epsilon\,\alpha_3\delta^2 q_X^* q + i\alpha_1\delta^2|q|^2 & \epsilon\,\alpha_3\delta q_{XX} + i\alpha_1\delta q_X + 2\epsilon\delta^3\alpha_3\,|q|^2 q \\ i\alpha_1\delta q_X^* - \epsilon\,\alpha_3\delta q_{XX}^* - 2\epsilon\delta^3\alpha_3\,|q|^2 q^* & \epsilon\,\alpha_3\delta^2 q_X^* q - \epsilon\,\alpha_3\delta^2 q_X q^* - i\alpha_1\delta^2|q|^2 \end{pmatrix},$$

where λ is a spectral parameter and $\zeta(T, X, \lambda)$ is a vector or matrix function.

When $\alpha_2 = -2\delta^2\alpha_1, \alpha_4 = -6\delta^2\alpha_3$ and $\alpha_5 = 6\delta^2\alpha_3$, Eq. (4) becomes the dark soliton version of Hirota equation

$$q_T = i\alpha_1 q_{XX} - 2i\delta^2\alpha_1|q|^2 q + \epsilon\,\alpha_3 q_{XXX} - 6\epsilon\,\delta^2\alpha_3|q|^2 q_X, \tag{18}$$

with linear spectral problem Eq. (15) and

$$F = \begin{pmatrix} -i\lambda & \delta q \\ \delta q^* & i\lambda \end{pmatrix}, \tag{19}$$

$$G = 4i\epsilon\,\lambda^3\alpha_3 \begin{pmatrix} 1 & 0 \\ 0 & -1 \end{pmatrix} - 2\lambda^2 \begin{pmatrix} i\alpha_1 & 2\epsilon\,\alpha_3\delta\,q \\ 2\epsilon\,\alpha_3\delta q^* & -i\alpha_1 \end{pmatrix}$$

$$+ 2\lambda \begin{pmatrix} i\epsilon\,\alpha_3\delta^2|q|^2 & -i\epsilon\,\alpha_3\delta q_X + \alpha_1\delta q \\ \alpha_1\delta q^* + i\epsilon\,\alpha_3\delta q_X^* & -i\epsilon\,\alpha_3\delta^2|q|^2 \end{pmatrix} \tag{20}$$

$$+ \begin{pmatrix} \epsilon\,\alpha_3\delta^2 q_X^* q - \epsilon\,\alpha_3\delta^2 q_X q^* - i\alpha_1\delta^2|q|^2 & \epsilon\,\alpha_3\delta q_{XX} + i\alpha_1\delta q_X - 2\epsilon\,\delta^3\alpha_3|q|^2 q \\ \epsilon\,\alpha_3\delta q_{XX}^* - i\alpha_1\delta q_X^* - 2\epsilon\,\delta^3\alpha_3|q|^2 q^* & \epsilon\,\alpha_3\delta^2 q_X q^* - \epsilon\,\alpha_3\delta^2 q_X^* q + i\alpha_1\delta^2|q|^2 \end{pmatrix}.$$

4. The bright soliton solutions for Eq. (14)

In this section, we propose the N-bright soliton solutions of Eq. (14) using the Riemann-Hilbert formulation (24–28). Let us consider Eq. (14) for localized solutions, i.e. assuming that potential function q decay to zero sufficiently fast as $X, T \to \pm\infty$. In the Riemann-Hilbert formulation, we treat ζ as a fundamental matrix of the two linear equations in (15). From (15) we note that when $X, T \to \pm\infty$, one has $\zeta = e^{-i\lambda\Lambda X + (4i\alpha_3\epsilon\lambda^3 - 2i\lambda^2\alpha_1)\Lambda T}$ with $\Lambda = \mathrm{diag}(1, -1)$. This motivates us to introduce the variable transformation

$$\zeta = J e^{-i\lambda\Lambda X + (4i\alpha_3\epsilon\lambda^3 - 2i\lambda^2\alpha_1)\Lambda T}, \tag{21}$$

where J is (X, T)-independent at infinity. Inserting (21) into (15) with (16)-(17), we get

$$J_X = -i\lambda[\Lambda, J] + \delta Q J, \tag{22a}$$
$$J_T = -(2i\alpha_1\lambda^2 - 4i\alpha_3\epsilon\lambda^3)[\Lambda, J] + VJ, \tag{22b}$$

with

$$Q = \begin{pmatrix} 0 & q \\ -q^* & 0 \end{pmatrix}, \quad V = (2\lambda\alpha_1\delta - 4\lambda^2\epsilon\alpha_3\delta)Q + 2\lambda \begin{pmatrix} -i\epsilon\alpha_3\delta^2|q|^2 & -i\epsilon\alpha_3\delta q_X \\ -i\epsilon\alpha_3\delta q_X^* & i\epsilon\alpha_3\delta^2|q|^2 \end{pmatrix}$$
$$+ \begin{pmatrix} \epsilon\alpha_3\delta^2 q_X q^* - \epsilon\alpha_3\delta^2 q_X^* q + i\alpha_1\delta^2|q|^2 & \epsilon\alpha_3\delta q_{XX} + i\alpha_1\delta q_X + 2\epsilon\delta^3\alpha_3|q|^2 q \\ i\alpha_1\delta q_X^* - \epsilon\alpha_3\delta q_{XX}^* - 2\epsilon\delta^3\alpha_3|q|^2 q^* & \epsilon\alpha_3\delta^2 q_X^* q - \epsilon\alpha_3\delta^2 q_X q^* - i\alpha_1\delta^2|q|^2 \end{pmatrix}.$$

Here $[\Lambda, J] = \Lambda J - J\Lambda$ is the commutator, $\mathrm{tr}(Q) = \mathrm{tr}(V) = 0$ and

$$Q^\dagger = -Q, \quad V^\dagger = -V, \tag{23}$$

where † represents the Hermitian of a matrix.

In what folows, we consider the scattering problem of the Eq. (22a). By doing so, the variable T is fixed and is a dummy variable. We first introduce the matrix Jost solutions $J_\pm(X, \lambda)$ of (22a) with the asymptotic condition

$$J_\pm \to I, \quad \text{when } X \to \pm\infty, \tag{24}$$

where I is a 2×2 unit matrix. Here the subscripts in $J_\pm$ refer to which end of the X-axis the boundary conditions are set. Then due to $\mathrm{tr}(Q) = 0$ and Abel's formula we have $\det(J_\pm) = 1$ for all X. Next we denote $E = e^{-i\lambda\Lambda X}$. Since $\Psi \equiv J_+E$ and $\Phi \equiv J_-E$ are both solutions of the first equation in (15), they must be linearly related, i.e.

$$J_-E = J_+ES(\lambda), \quad \lambda \in \mathbb{R} \tag{25}$$

where

$$S(\lambda) = \begin{pmatrix} s_{11} & s_{12} \\ s_{21} & s_{22} \end{pmatrix}, \quad \lambda \in \mathbb{R}$$

is the scattering matrix, and $\mathbb{R}$ is the set of real numbers. Notice that $\det(S(\lambda)) = 1$ since $\det(J_\pm) = 1$. If we denote (Φ, Ψ) as a collection of columns,

$$\Phi = [\phi_1, \phi_2], \quad \Psi = [\psi_1, \psi_2], \tag{26}$$

by using the same formulation as (24; 25; 27), we have the Jost solution

$$P^+ = [\phi_1, \psi_2]e^{i\lambda\Lambda X} = J_-H_1 + J_+H_2, \tag{27}$$

is analytic in $\lambda \in C_+$, and Jost solution

$$P^- = e^{-i\lambda\Lambda X} \begin{bmatrix} \hat{\phi}_1 \\ \hat{\psi}_2 \end{bmatrix} = H_1 J_-^{-1} + H_2 J_+^{-1}, \tag{28}$$

is analytic in $\lambda \in C_-$, with

$$\Phi^{-1} = \begin{bmatrix} \hat{\phi}_1 \\ \hat{\phi}_2 \end{bmatrix}, \quad \Psi^{-1} = \begin{bmatrix} \hat{\psi}_1 \\ \hat{\psi}_2 \end{bmatrix},$$

and

$$H_1 = \text{diag}(1,0), \quad H_2 = \text{diag}(0,1).$$

In addition, it is easy to see that

$$P^+(X,\lambda) \to I, \quad \text{as } \lambda \in C_+ \to \infty, \tag{29}$$

and

$$P^-(X,\lambda) \to I, \quad \text{as } \lambda \in C_- \to \infty. \tag{30}$$

In addition, if we express S^{-1} as

$$S^{-1} = \begin{pmatrix} \hat{s}_{11} & \hat{s}_{12} \\ \hat{s}_{21} & \hat{s}_{22} \end{pmatrix}, \quad \lambda \in \mathbb{R},$$

from $\det(S(\lambda)) = 1$ we have

$$\hat{s}_{11} = s_{22}, \quad \hat{s}_{22} = s_{11}, \quad \hat{s}_{12} = -s_{12}, \quad \hat{s}_{21} = -s_{21}. \tag{31}$$

Hence we have constructed two matrix functions P^+ and P^- which are analytic in C_+ and C_-, respectively. On the real line, using Eqs. (25), (27) and (28), it is easily to see that

$$P^-(X,\lambda)P^+(X,\lambda) = G(X,\lambda), \quad \lambda \in \mathbb{R}, \tag{32}$$

with

$$G = E(H_1 + H_2S)(H_1 + S^{-1}H_2)E^{-1} = E \begin{pmatrix} 1 & \hat{s}_{12} \\ s_{21} & 1 \end{pmatrix} E^{-1}.$$

This determines a matrix Riemann-Hilbert problem with asymptotics

$$P^\pm(X,\lambda) \to I, \quad \text{as } \lambda \to \infty, \tag{33}$$

which provide the canonical normalization condition for this Riemann-Hilbert problem. If this problem can be solved, one can readily reconstruct the potential $q(X,T)$ as follows. Notice that P^+ is the solution of the spectral problem (22a). Thus if we expand P^+ at large λ as

$$P^+(X,\lambda) = I + \frac{1}{\lambda}P_1^+(X) + O(\lambda^{-2}), \quad \lambda \to \infty, \tag{34}$$

and inserting this expansion into (22a), then comparing $O(1)$ terms in (34), we find that

$$\delta Q = i[\Lambda, P_1^+] = \begin{pmatrix} 0 & 2iP_{12} \\ -2iP_{21} & 0 \end{pmatrix}. \tag{35}$$

Thus, recalling the definition of Q the potentials q is reconstructed immediately as

$$q = 2iP_{12}/\delta,\tag{36}$$

where $P_1^+ = (P_{ij})$. In addition, from the definitions of P^+, P^- and Eq. (25) we have

$$\det P^+ = \hat{s}_{22} = s_{11}, \quad \det P^- = s_{22} = \hat{s}_{11}.\tag{37}$$

The symmetry properties of the potential Q and V in (23) give rise to symmetry properties in the scattering matrix as well as in the Jost functions. In fact, after some computation we have $J_\pm$ satisfies the involution property

$$J_\pm^\dagger(X, \lambda^*) = J_\pm^{-1}(X, \lambda),\tag{38}$$

analytic solutions $P^\pm$ satisfy the involution property

$$(P^+)^\dagger(\lambda^*) = P^-(\lambda),\tag{39}$$

and S satisfies the involution property

$$S^\dagger(\lambda^*) = S^{-1}(\lambda).\tag{40}$$

Let λ_k and $\bar{\lambda}_k$ are zero points of $\det P^+$ and $\det P^-$, respectively. We see from (37) that $(\lambda_k, \bar{\lambda}_k)$ are zeros of the scattering coefficients $\hat{s}_{22}(\lambda)$ and $s_{22}(\lambda)$. Due to the above involution property, we have the symmetry relation

$$\bar{\lambda}_k = \lambda_k^*.\tag{41}$$

For simplicity, we assume that all zeros $\{(\lambda_k, \bar{\lambda}_k), k = 1, 2, \cdot, N\}$ are simple zeros of $\hat{s}_{22}(\lambda)$ and $s_{22}(\lambda)$, then each kernal of $P^+(\lambda_k)$ and $P^-(\bar{\lambda}_k)$ contains only a single column vector v_k and row vector $\bar{v}_k$,

$$P^+(\lambda_k)v_k = 0, \quad \bar{v}_k P^-(\bar{\lambda}_k) = 0.$$

Taking the Hermitian of the above equations and using the involution properties, we have

$$\bar{v}_k = v_k^\dagger.\tag{42}$$

To obtain the soliton solutions, we set $G = I$ in (32). In this case, the solutions to this special Riemann-Hilbert problem have been derived in (25; 26) as

$$P_1^+(T, X, \lambda) = \sum_{j,k=1}^N v_j \left(M^{-1}\right)_{jk} \bar{v}_k,\tag{43}$$

where

$$M_{jk} = \frac{\bar{v}_j v_k}{\bar{\lambda}_j - \lambda_k}.\tag{44}$$

The zeros λ_k and $\bar{\lambda}_k$ are T-independent. To find the spatial and temporal evolutions for vectors $v_k(T, X)$, we take the X-derivative to equation $P^+ v_k = 0$. By using (22a), one gets

$$P^+(X, \lambda_k)\left(\frac{\partial v_k}{\partial X} + i\lambda_k \Lambda v_k\right) = 0,\tag{45}$$

thus we have

$$\frac{dv_k}{dX} + i\lambda_k \Lambda v_k = 0. \tag{46}$$

Similarly, taking T-derivative to equation $P^+v_k = 0$ and using (22b), one has

$$P^+(T, X, \lambda_k)\left(\frac{\partial v_k}{\partial T} + (2i\,\alpha_1\lambda_k{}^2 - 4\,i\alpha_3\epsilon\lambda_k{}^3)v_k\right) = 0, \tag{47}$$

thus we have

$$\frac{\partial v_k}{\partial T} + (2i\,\alpha_1\lambda_k{}^2 - 4\,i\alpha_3\epsilon\lambda_k{}^3)v_k = 0. \tag{48}$$

Solving (46) and (48) we get

$$v_k(T, X) = e^{-i\lambda_k\Lambda X + (4i\alpha_3\epsilon\lambda_k{}^3 - 2i\,\lambda_k{}^2\alpha_1)\Lambda T}v_{k0}, \tag{49a}$$

$$\bar{v}_k(T, X) = \bar{v}_{k0}e^{i\lambda_k\Lambda X + (-4i\alpha_3\epsilon\lambda_k{}^3 + 2i\,\lambda_k{}^2\alpha_1)}, \tag{49b}$$

where $(v_{k0}, \bar{v}_{k0})$ are constant vectors.

In summary, the N-bright soliton solutions to Eq. (14) are obtained from the analytical functions P_1^+ in (43) together with the potential reconstruction formula (36) as

$$q(T, X) = 2iP_{12}/\delta = 2i\left(\sum_{j,k=1}^{N} v_j\left(M^{-1}\right)_{jk}\bar{v}_k\right)_{12}/\delta, \tag{50}$$

where the vectors v_j are given by (49). Without loss of generality, we take $v_{k0} = [b_k, 1]'$ with b_k constants. And if we denote

$$\xi_k = -i\lambda_k X + (4\,i\alpha_3\epsilon\lambda_k{}^3 - 2i\,\lambda_k{}^2\alpha_1)T, \tag{51}$$

the general N-soliton solution to Eq. (14) can be written out explicitly as

$$q(T, X) = \frac{2i}{\delta}\sum_{j,k=1}^{N} b_j e^{\xi_j - \xi_k^*}(M^{-1})_{jk}, \tag{52}$$

with

$$M_{jk} = \frac{1}{\lambda_j^* - \lambda_k}\left(b_j^* c_k e^{\xi_k + \xi_j^*} - e^{-\xi_k - \xi_j^*}\right). \tag{53}$$

In what follows, we investigate the dynamics of the one-soliton and two-soliton solutions in Eqs. (14) in detail.

4.1 Examples of single and two bright solitons in Eq. (14)

To get the single bright soliton solution for Eq. (14), we set $N = 1$ in (52) to have

$$q(T, X) = \frac{2i(\lambda_1^* - \lambda_1)}{\delta}\frac{b_1 e^{\xi_1 - \xi_1^*}}{e^{-\xi_1 - \xi_1^*} + |b_1|^2 e^{\xi_1 + \xi_1^*}}. \tag{54}$$

If setting $\lambda_1 = \zeta_1 + i\eta_1$, $b_1 = e^{-2\eta_1 X_0 + i\omega_0}$, the single soliton solution (54) can be rewritten as

$$q(T, X) = \frac{2\eta_1}{\delta}\text{sech}[2\,\eta_1(X + \left(4\,\alpha_3\epsilon\,\eta_1{}^2 + 4\alpha_1\zeta_1 - 12\,\alpha_3\epsilon\,\zeta_1{}^2\right)T - X_0)]\exp^{i\theta}, \tag{55}$$

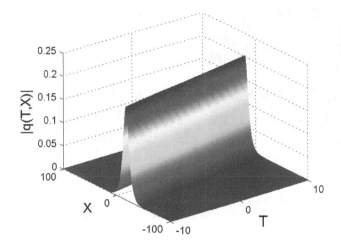

Fig. 1. (color online). Evolution of single soliton $|q(T, X)|$ in (55) with parameters (56). It is similar to single soliton in standard NLS equation.

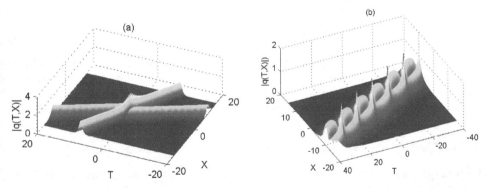

Fig. 2. (color online). The shapes of two-soliton solutions $|q(T, X)|$ in (52) with (53). (a) soliton collision with parameters (57); (b) bound state with parameters (58).

with $\theta = -2\zeta_1 X + \left(-4\alpha_1\zeta_1^2 + 4\alpha_1\eta_1^2 + 8\alpha_3\epsilon\zeta_1^3 - 24\alpha_3\epsilon\zeta_1\eta_1^2\right) T + \omega_0$, and X_0, ω_0 are constants. This solution is similar to the solitary wave solution in the standard NLS equation (1). Its amplitude function has the shape of a hyperbolic secant with peak amplitude $2\eta_1/\delta$, and its velocity depends on several parameters, which is $12\alpha_3\epsilon\zeta_1^2 - 4\alpha_3\epsilon\eta_1^2 - 4\alpha_1\zeta_1$. The phase θ of this solution depends linearly both on space X and time T. We show this single soliton solution in Fig. 1 with parameters

$$\zeta_1 = 0.5, \eta_1 = 0.1, X_0 = 1.5, \omega_0 = 2, \delta = 1, \alpha_1 = 0.5, \alpha_3 = 1, \epsilon = 1. \tag{56}$$

The two-soliton solution in Eq. (14) corresponds to $N = 2$ in the general N-soliton solution (52) with (53). This solution can also be written out explicitly, however, we prefer to showing

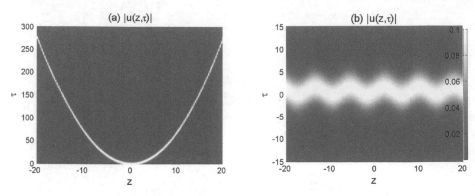

Fig. 3. (color online). Evolution of single soliton solutions $|u(z, \tau)|$ in HNLS equation (2) with controlable coefficients (59) and (62), respectively. (a) Soliton solution (61) with parameter (56) and $\rho_0 = 0.5, c = 1, k = 2$. (b) Soliton solution (64) with parameter (59) and $\rho_0 = 0.5, c = 1, k = 2$.

their behaviors by figures, see Fig. 2(a)-(b). Below we take $\lambda_1 = \zeta_1 + i\eta_1$ and $\lambda_2 = \zeta_2 + i\eta_2$ and examine this solution with various velocity parameters: one is $12\,\alpha_3\epsilon\,\zeta_1{}^2 - 4\,\alpha_3\epsilon\,\eta_1{}^2 - 4\,\alpha_1\zeta_1 = 12\,\alpha_3\epsilon\,\zeta_2{}^2 - 4\,\alpha_3\epsilon\,\eta_2{}^2 - 4\,\alpha_1\zeta_2$, i.e. the collision between two solitons, and the other is $12\,\alpha_3\epsilon\,\zeta_1{}^2 - 4\,\alpha_3\epsilon\,\eta_1{}^2 - 4\,\alpha_1\zeta_1 \neq 12\,\alpha_3\epsilon\,\zeta_2{}^2 - 4\,\alpha_3\epsilon\,\eta_2{}^2 - 4\,\alpha_1\zeta_2$, i.e. bound state. In Fig. 2(a), the two soliton parameters in Eq. (52) with (53) are

$$\alpha_1 = 0.5, \ \alpha_3 = 0.8, \ \epsilon = 1, \ \delta = 1, \ \lambda_1 = 0.2 + 0.7i, \ \lambda_2 = -0.1 + 0.5i, \ b_1 = 1, \ b_2 = 1. \quad (57)$$

Under these parameters, the velocity of the two solitons are different. It is observed that interactions between two soliton don't change the shape and velocity of the solitons, and there is no energy radiation emitted to the far field. Thus the interaction of these solitons is elastic, which is a remarkable property which signals that the HNLS equation (14) is integrable. Fig. 2(b) displays a bound state in Eq. (14), and the soliton parameters here are

$$\alpha_1 = 0.5, \ \alpha_3 = 0.8, \ \epsilon = 1, \ \delta = 1, \ \lambda_1 = 0.3i, \ \lambda_2 = -0.1 + 0.4272i, \ b_1 = 1, \ b_2 = 1. \quad (58)$$

Under these parameters, the two constituent solitons have equal velocities, thus they will stay together to form a bound state which moves at the common speed. It can be seen that the width of this solution changes periodically with time, thus this solution is called breather soliton.

5. Dynamics of solitons in HNLS equation (2)

In what follows, we investigate the dynamic behavior of solitons in the variable-coefficients HNLS equation (2) with special soliton management parameters $d_j (j = 1, 2, 3, 4, 5, 6)$.

5.1 Single soliton solutions

We choose two cases of soliton management parameters $d_j (j = 1, 2, 3, 4, 5, 6)$ to study the dynamics of the single solitons in HNLS equation (2). Firstly, if we take the soliton management parameters to satisfy

$$d_1 = 1.6 \ (k\alpha_1 - 3\,\alpha_3 c)\,z/k^3, \ d_2 = 1.6 \ (\alpha_2 k - \alpha_4 c)\,z/\rho_0{}^2 k,$$

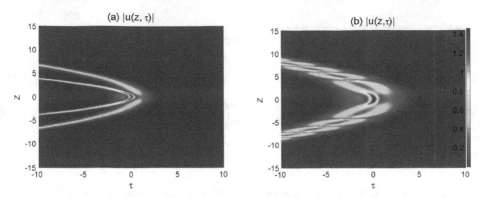

Fig. 4. (color online). The two-soliton solutions $|u(z,\tau)|$ in HNLS equation (2) with coefficients (59). (a) soliton collision with parameter (57) and $\rho_0 = 0.5, c = 1, k = 2$; (b) bound state with parameter (58) and $\rho_0 = 0.5, c = 1, k = 2$.

$$d_3 = 1.6\,\alpha_3 z/k^3, \ d_4 = 1.6\,\alpha_4 z/\rho_0{}^2 k, \ d_5 = 1.6\,\alpha_5 z/\rho_0{}^2 k, \ d_6 = 0, \tag{59}$$

the variables ρ, T and X in similarity transformation (3) are

$$\rho = \rho_0, \ T = 0.8z^2, \ X = k\tau + (2.4\,c^2\alpha_3 - 1.6\,ck\alpha_1)z^2/k^2. \tag{60}$$

So the single soliton solution in HNLS equation (2) with coefficients (59) is

$$u(z,\tau) = \rho_0 q\,(T,X)\,e^{ic(k^3\tau + 1.6\,c^2\alpha_3\,z^2 - 0.8\,ck\alpha_1 z^2)/k^3}, \tag{61}$$

where $q\,(T,X)$ satisfies Eq. (55) and T, X satisfy Eq. (60).
Secondly, if we take the soliton management parameters to satisfy

$$d_1 = 0.8\cos(0.8z)\,(k\alpha_1 - 3\,\alpha_3 c)\,/k^3, d_2 = 0.8\cos(0.8z)\,(\alpha_2 k - \alpha_4 c)\,/\rho_0{}^2 k, d_6 = 0,$$

$$d_3 = 0.8\,\alpha_3\cos(0.8z)\,/k^3, \ d_4 = 0.8\,\alpha_4\cos(0.8z)\,/\rho_0{}^2 k, \ d_5 = 0.8\,\alpha_5\cos(0.8z)\,/\rho_0{}^2 k, \tag{62}$$

the variables ρ, T and X in similarity transformation (3) are

$$\rho = \rho_0, \ T = \sin(0.8z), \ X = k\tau + (3\,c^2\alpha_3 - 2\,ck\alpha_1)\sin(0.8z)\,/k^2. \tag{63}$$

In this case the single soliton solution in HNLS equation (2) with coefficients (62) is

$$u(z,\tau) = \rho_0 q\,(T,X)\,e^{ic(k^3\tau + 2c^2\alpha_3\sin(0.8z) - ck\alpha_1\sin(0.8z))/k^3}, \tag{64}$$

where $q\,(T,X)$ satisfies Eq. (55) and T, X satisfy Eq. (63).
In Fig. 3, we show the single soliton solutions (61) and (64) in HNLS equation (2) with coefficients (59) and (62), respectively. Here the solution parameters are given in (56) and $\rho_0 = 0.5, c = 1, k = 2$. It is observered that when the soliton management parameters $d_j (j = 1, 2, 3, 4, 5)$ are linearly dependent on variable z and $d_6 = 0$ (see Eq. (59)), the trajectory of the optical soliton is a localized parabolic curve, as shown in Fig. 3(a). When the soliton management parameters $d_j (j = 1, 2, 3, 4, 5)$ are periodically dependent on variable z and $d_6 = 0$ (see Eq. (62)), the trajectory of the optical soliton is a periodical localized nonlinear wave, as shown in Fig. 3(b).

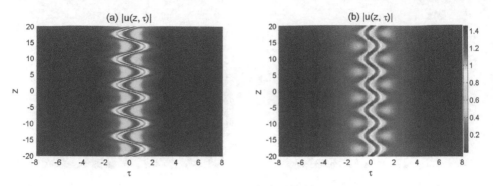

Fig. 5. (color online). The two-soliton solutions $|u(z,\tau)|$ in HNLS equation (2) with coefficients (62). (a) soliton collision with parameter (57) and $\rho_0 = 0.5, c = 1, k = 2$; (b) bound state with parameter (58) and $\rho_0 = 0.5, c = 1, k = 2$.

5.2 Collisions of the two-solitons

We now demonstrate various collision scenarios in HNLS equation (2) with coefficients (59) and (62), respectively. As in Section 4.1, we consider the two-soliton collisions and bound states in equation (2).

When the coefficients of equation (2) satisfies (59), its two-soliton solution is

$$u(z,\tau) = \rho_0 q\,(T,X)\,e^{ic(k^3\tau + 1.6\,c^2\alpha_3\,z^2 - 0.8\,ck\alpha_1 z^2)/k^3}, \tag{65}$$

where T, X satisfy Eq. (60), and $q\,(T,X)$ satisfies Eq. (52) with (53) and $N = 2$.

When the coefficients of equation (2) satisfies (62), its two-soliton solution is

$$u(z,\tau) = \rho_0 q\,(T,X)\,e^{ic(k^3\tau + 2\,c^2\alpha_3\sin(0.8z) - ck\alpha_1\sin(0.8z))/k^3}, \tag{66}$$

where T, X satisfy Eq. (63), and $q\,(T,X)$ satisfies Eq. (52) with (53) and $N = 2$.

In Fig. 4, we display the evolutions of the two-soliton solutions (65) in HNLS equation (2) with coefficients (59). Fig. 4(a) shows the soliton collision with parameter (57) and $\rho_0 = 0.5, c = 1, k = 2$, and Fig. 4(b) shows the bound state with parameter (58) and $\rho_0 = 0.5, c = 1, k = 2$. In Fig. 5, we display the evolutions of the two-soliton solutions (66) in HNLS equation (2) with coefficients (62). Fig. 5(a) shows the soliton collision with parameter (57) and $\rho_0 = 0.5, c = 1, k = 2$, and Fig. 5(b) shows the bound state with parameter (58) and $\rho_0 = 0.5, c = 1, k = 2$.

6. Conclusions

In summary, we have studied the variable-coefficient higher order nonlinear Schrödinger equation which describes the wave propagation in a nonlinear fiber medium with higher-order effects such as third order dispersion, self-steepening and stimulated Raman scattering. By means of similarity transformation, we first change this variable-coefficient equation into the constant-coefficient HNLS equation. Then we investigate the integrability of the constant-coefficient HNLS equation by prolongation technique and find two Lax integrable HNLS equations. The exact bright N-soliton solutions for the bright soliton version of HNLS equation are obtained using Riemann-Hilbert formulation. Finally, the dynamics of the optical solitons in both constant-coefficient and variable-coefficient HNLS equations is examined and the effects of higher-order effects on the velocity and shape of the optical soliton are observed. In addition, it is seen that the bright two-soliton solution of the HNLS equation behaves in an elastic manner characteristic of all soliton solutions.

7. Acknowledgments

This work was supported by NSFC under grant No. 11001263 and China Postdoctoral Science Foundation.

8. References

[1] K. E. Lonngren, Soliton experiments in plasmas, Plas. Phys. 25, 943 (1983).

[2] G.P. Agrawal, Nonlinear Fiber Optics, Academic Press, San Diego (1989).

[3] A. Hasegawa and Y. Kodama, Solitons in Optical Communications, Clarendon, Oxford, 1995.

[4] J. Denschlag, J. E. Simsarian, D. L. Feder, Charles W. Clark, L. A. Collins, J. Cubizolles1, L. Deng, E. W. Hagley, K. Helmerson, W. P. Reinhardt1, S. L. Rolston, B. I. Schneider and W. D. Phillips, Generating Solitons by Phase Engineering of a Bose-Einstein Condensate, Science, 287, 97-101 (2000).

[5] B. P. Anderson, P. C. Haljan, C. A. Regal, D. L. Feder, L. A. Collins, C. W. Clark, and E. A. Cornell, Watching Dark Solitons Decay into Vortex Rings in a Bose-Einstein Condensate, Phys. Rev. Lett. 86, 2926-2929 (2001).

[6] M. A. Helal, Soliton solution of some nonlinear partial differential equations and its applications in fluid mechanics, Chaos, Solitons and Fractals, 13, 1917-1929 (2002).

[7] A. Hasegawa and F. Tappert, Transmission of stationary nonlinear optical pulses in dispersive dielectric fibers. I. Anomalous dispersion, Appl. Phys. Lett. 23, 142-144 (1973); Transmission of stationary nonlinear optical pulses in dispersive dielectric fibers. II. Normal dispersion, Appl. Phys. Lett. 23, 171-172 (1973).

[8] L. F. Mollenauer, R. H. Stolen, and J. P. Gordon, Experimental observation of picosecond pulse narrowing and solitons in optical fibers, Phys. Rev. Lett. 45, 1095-1098 (1980).

[9] V. E. Zakharov and A. B. Shabat, Exact theory of two-dimensional self-focusing and one-dimensional self-modulation of manes in nonlinear media, Zh. Eksp. Teor. Fiz. 61, 118 (1971) (Sov. Phys. JETP 34, 62, 1972).

[10] V. E. Zakharov and A. B. Shabat, On interactions between solitons in a stable medium, Zh. Eksp. Teor. Fiz. 64, 1627 (1973) (Sov. Phys. JETP 37, 823, 1974).

[11] J. Yang, Nonlinear waves in integrable and non-integrable systems, Society for Industrial and Applied Mathematics, U.S. (2010).

[12] Y. Kodama and A. Hasegawa, Nonlinear pulse propagation in a monomode dielectric guide, IEEE J. Quantum Electron. 23, 510 (1987).

[13] K. Porsezian and K. Nakkeeran, Optical Solitons in Presence of Kerr Dispersion and Self-Frequency Shift, Phys. Rev. Lett. 76, 3955 (1996).

[14] K. Nakkeeran, K. Porsezian, P. Shanmugha Sundaram, and A. Mahalingam, Optical solitons in N-coupled higher order nonlinear Schrodinger equations, Phys. Rev. Lett. 80, 1425 (1998).

[15] E. Papaioannou, D. J. Frantzeskakis, K. Hizanidis, An analytical treatment of the effect of axial inhomogeneity on femtosecond solitary waves near the zero dispersion point, 32, 145-154 (1996).

[16] A. Mahalingam and K. Porsezian, Propagation of dark solitons with higher-order effects in optical fibers, Phys. Rev. E 64, 046608 (2001).

[17] Z. Y. Xu, L. Li, Z. H. Li, and G. S. Zhou, Modulation instability and solitons on a cw background in an optical fiber with higher-order effects, Phys. Rev. E 67, 026603 (2003).

[18] D. S. Wang, X. H. Hu, J. P. Hu, and W. M. Liu, Quantized quasi-two-dimensional Bose-Einstein condensates with spatially modulated nonlinearity, Phys. Rev. A 81, 025604 (2010).

[19] D. S. Wang, X. H. Hu, and W. M. Liu, Localized nonlinear matter waves in two-component Bose-Einstein condensates with time- and space-modulated nonlinearities, Phys. Rev. A 82, 023612 (2010).

[20] H.D. Wahlquist and F.B. Estabrook, Prolongation structures of nonlinear evolution equations. J. Math. Phys. 16, 1-7 (1975).

[21] F.B. Estabrook and H.D. Wahlquist, Prolongation structures of nonlinear evolution equations II, J.Math.Phys.17, 1293-1297 (1976).

[22] D. S. Wang, Integrability of the coupled KdV equations derived from two-layer fluids: Prolongation structures and Miura transformations, Nonlinear Analysis, 73, 270-281 (2010). Deng-Shan Wang

[23] R. Dodd and A. P. Fordy, The prolongation structure of quasi-polynomial flows Proc. R. Soc. London A 385, 389-429 (1983).

[24] V.S. Gerdjikov, Basic aspects of soliton theory, Sixth International Conference on Geometry, Integrability and Quantization June 3-10, 2004, Varna, Bulgaria Ivailo M. Mladenov and Allen C. Hirshfeld, Editors, Sofia 1-48 (2005).

[25] V. E. Zakharov, S. V. Manakov, S. P. Novikov, and L. P. Pitaevskii, *The Theory of Solitons: The Inverse Scattering Method*, Consultants Bureau, New York, 1984.

[26] V.S. Shchesnovich and J. Yang, General soliton matrices in the Riemann-Hilbert problem for integrable nonlinear equations. J. Math. Phys. 44, 4604 (2003).

[27] E.V. Doktorov, J. Wang, and J. Yang, Perturbation theory for bright spinor Bose-Einstein condensate solitons, Phys. Rev. A. 77, 043617 (2008).

[28] D. S. Wang, D. J. Zhang and J. Yang, Integrable properties of the general coupled nonlinear Schrodinger equations, J. Math. Phys. 51, 023510 (2010).

3

Spontaneous Nonlinear Scattering Processes in Silica Optical Fibers

Edouard Brainis
Université libre de Bruxelles
Belgium

1. Introduction

When light travels in a optical fiber, a fraction of its total power is always scattered to other wavelengths (or polarization) due to material non linearity. Whether that scattering is weak or strong, desirable or not, depends on the situation. One distinguishes (i) scattering stimulated by the presence of a seed wave (at another wavelength or polarization), (ii) spontaneous scattering, and (iii) amplified spontaneous scattering. Stimulated Raman scattering (SRS), stimulated Brillouin scattering (SBS) and four-wave mixing (FWM) are examples of stimulated scatterings. Those have been thoroughly studied in the past thirty years and are well summarized in classic nonlinear fiber optics textbooks, e.g. (Agrawal, 2007). Several chapters of this book also deal with specific aspects and applications of stimulated scattering. The present chapter focuses on *spontaneous* scattering processes, cases (ii) and (iii).

The chapter also concentrates on nonlinear scattering in *silica* fibers because nowadays those are the most common and widely used types of fibers. Gas-filled hollow core fibers (Benabid et al., 2005) and ion doped fibers (Digonnet, 2001) are not considered here, and it is assumed that the fiber has not been subjected to poling (Bonfrate et al., 1999; Huy et al., 2007; Kazansky et al., 1997), so that the main non linearity is of third order. In this context the most important spontaneous nonlinear scattering processes are

1. the spontaneous Raman scattering (RS),
2. the spontaneous Brillouin scattering (BS), and
3. the spontaneous four-photon scattering (FPS).

These phenomena play an important role in many applications of optical fibers. This role can be positive as in remote optical sensing (Alahbabi et al., 2005a;b; Dakin et al., 1985; Farahani & Gogolla, 1999; Wait et al., 1997). It can also be detrimental as in fiber optics telecommunication, where spontaneous nonlinear scattering processes contribute to decrease the signal-to-noise ratio (SNR) or in supercontinuum generation, where it limits the coherence and stability of the supercontinuum (Corwin et al., 2003; Dudley et al., 2006). In the emerging field of quantum photonics, fiber optical photon-pair sources are intrinsically based on the physics of the FPS (Amans et al., 2005; Brainis, 2009; Brainis et al., 2005), while at the same time RS is the main factor that limits the SNR (Brainis et al., 2007; Dyer et al., 2008; Fan & Migdall, 2007; Lee et al., 2006; Li et al., 2004; Lin et al., 2006; 2007; Rarity et al., 2005; Takesue, 2006).

This chapter reviews the physics of spontaneous nonlinear scattering processes in optical fibers. In Sec. 2, the physical origin of RS, BS and FPS in explained. Because those are pure quantum mechanical effects, they cannot be properly described in the framework of classical nonlinear optics. A quantum mechanical treatment is presented in Sec. 3. Finally, in Sec. 4, the coupling between different scattering processes in considered.

2. Physics of nonlinear scattering processes in optical fibers

2.1 Raman scattering

Light at frequency ω_p traveling in an optical fiber, can excite the fiber molecules from ground to excited vibrational states. In amorphous silica fiber, vibrational states have energy $\hbar|\Omega|$ with $|\Omega|/(2\pi)$ in the 0-40 THz range.These energies (about 0.05 eV) being much smaller than the photon energy $\hbar\omega_p$, no direct excitation of the vibrational states is possible. However, the states can be excited through a second order *Raman transition* involving a second photon at frequency ω_s and a virtual state as shown in Fig. 1. The spontaneous inelastic scattering that converts a ω_p photon into a $\omega_s = \omega_p - |\Omega|$ photon and a vibrational excitation at frequency $|\Omega|$ is call a spontaneous *Stokes* process. If a vibrational state at frequency $|\Omega|$ is initially populated, the complementary process in which a ω_p photon is converted into a $\omega_a = \omega_p + |\Omega|$ photon is also allowed and called a spontaneous *anti-Stokes* process, see Fig. 1.

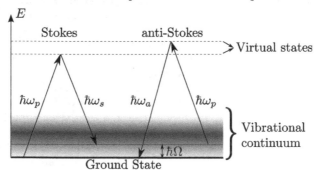

Fig. 1. Spontaneous Stokes and anti-Stokes processes in amorphous silica fibers

Molecular vibrations behave like waves (phonons). The momentum of these vibrational waves corresponds to the momentum mismatch of the pump and (anti-)Stoke waves and does not depend on $|\Omega|$. For this reason, Raman scattering has no preferential direction. It happens in the forward but also in the backward direction. The damping of a phonon wave depends on the wave number and is stronger for shorter wavelength. In fibers the damping is very strong because of the amorphous nature of silica. Therefore the molecular vibration can to a good approximation be considered as local. Yet the small difference in the forward and backward damping explains that the strengths of Raman scattering in forward and backward directions is slightly different (Bloembergen & Shen, 1964).

In addition to the Stokes and anti-Stokes processes that convert pump photons to other wavelengths, Raman scattering can also convert the Stokes and anti-Stokes photons at ω_s and ω_a back to the pump mode through *reverse* Stokes and anti-Stokes scattering. In Sec. 3.1, both direct and reverse scattering processes are taken into account to derive the basic equations governing the *net* energy transfer from the pump to Stokes and anti-Stokes waves. For a single monochromatic pump wave at ω_p the scattered spectral power density $S(z,\omega)$ obeys the following propagation equation

$$\frac{d}{dz}\mathcal{S}(z,\omega) = \left[\mathcal{S}(z,\omega)g(\omega_p,\Omega,\theta) + \frac{\hbar\omega}{2\pi}\left[m_{th}(|\Omega|) + \nu(\Omega)\right]|g(\omega_p,\Omega,\theta)|\right]P_p(z) \qquad (1)$$

where $\Omega = \omega_p - \omega$ (positive for a Stokes process and negative for an anti-Stokes ones), $\nu(\Omega)$ is the Heaviside step function, and

$$m_{th}(|\Omega|) = \left[\exp\left(\frac{\hbar|\Omega|}{k_BT}\right) - 1\right]^{-1} \qquad (2)$$

is the thermal equilibrium expectation value of the number of vibrational excitations at angular frequency $|\Omega|$. The function $g(\omega_p,\Omega,\theta)$ in Eq. (1) is the Raman *gain*. The Raman gain measures the scattering strength and is polarization dependent. For a linearly polarized pump field, the Raman gain is maximal for photon scattered with polarization parallel to the pump and minimal for photons scattered with polarization orthogonal to the pump (Stolen, 1979):

$$g(\omega_p,\Omega,\theta) = g_{\parallel}(\omega_p,\Omega)\cos^2(\theta) + g_{\perp}(\omega_p,\Omega)\sin^2(\theta), \qquad (3)$$

where θ is the angle between the linear polarization vectors of pump and scattered photons. The parallel and orthogonal gains are $g_{\parallel}(\omega_p,\Omega)$ and $g_{\perp}(\omega_p,\Omega)$ are material properties that can be measured experimentally. It can be shown (see Sec. 3.1) that the ratio of the Stokes to anti-Stokes gain corresponding to the same vibrational mode $|\Omega|$ is

$$\frac{g(\omega_p,\Omega,\theta)}{g(\omega_p,-\Omega,\theta)} = -\frac{n(\omega_p+|\Omega|)}{n(\omega_p-|\Omega|}\left(\frac{\omega_p-|\Omega|}{\omega_p+|\Omega|}\right)^3. \qquad (4)$$

Stokes and anti-Stokes gain have opposite signs: Stokes gain is positive while anti-Stokes gain is negative.

2.1.1 Spontaneous scattering

With initial condition $\mathcal{S}(0,\omega) = 0$, $\forall\omega \neq \omega_p$, Eq. (1) describes both spontaneous Raman scattering and its subsequent amplification. In the initial propagation stage, the first term in the square bracket can be neglected. This regime corresponds to pure spontaneous Raman scattering. The solution of Eq. (1) is

$$\mathcal{S}(L,\omega) = \frac{\hbar\omega}{2\pi}\left[m_{th}(|\Omega|) + \nu(\Omega)\right]|g(\omega_p,\Omega,\theta)|P_p L, \qquad (5)$$

where L is the propagation length. The strength of the spontaneous Raman parallel and orthogonal scattering is often measured by the parallel and orthogonal *spontaneous Raman coefficients*

$$R_{\parallel,\perp}(\omega_p,\Omega,T) = \frac{\hbar\omega_p}{2\pi}\left[m_{th}(|\Omega|) + \nu(\Omega)\right]|g_{\parallel,\perp}(\omega_p,\Omega)|. \qquad (6)$$

Spontaneous Raman scattering has been observed and measured in bulk glass (Hellwarth et al., 1975; Stolen & Ippen, 1973) and in optical fibers (Stolen et al., 1984; Wardle, 1999). In optical fiber, the polarization properties are usually more difficult to measure because standard fibers do not preserve and even scramble polarization. For this reason, the *effective spontaneous Raman coefficient* is often taken to be $R = (R_{\parallel} + R_{\perp})/2$.

It is interesting to note that the ratio of Stoke to anti-Stokes spectral components only depend on temperature:

$$\frac{\mathcal{S}(L,\omega_p - |\Omega|)}{\mathcal{S}(L,\omega_p + |\Omega|)} = \frac{n(\omega_p + |\Omega|)}{n(\omega_p - |\Omega|)} \left(\frac{\omega_p - |\Omega|}{\omega_p + |\Omega|}\right)^4 \exp\left(\frac{\hbar|\Omega|}{k_B T}\right). \tag{7}$$

This is the reason why spontaneous Raman scattering is used for temperature sensing (Alahbabi et al., 2005a;b; Dakin et al., 1985; Farahani & Gogolla, 1999; Wait et al., 1997).

2.1.2 Amplified spontaneous scattering

According to Eq. (1), the spontaneous scattering regime ends as soon as $\mathcal{S}(z,\omega)$ becomes significant compared to $\frac{\hbar\omega}{2\pi}[m_{\text{th}}(|\Omega|) + \nu(\Omega)]$. At that point, the scattering becomes stimulated and the system enters the amplification regime. From Eq. (5), one sees that the amplification regime is reached when $g(\omega_p, \Omega, \theta) P_p L \approx 1$. For $\mathcal{S}(0,\omega) = 0$, the solution of Eq. (1) is

$$\mathcal{S}(L,\omega) = \frac{\hbar\omega}{2\pi}[m_{\text{th}}(|\Omega|) + \nu(\Omega)]\left|e^{g(\omega_p, \Omega, \theta) P_p L} - 1\right|. \tag{8}$$

Stokes radiation ($\Omega > 0$, $g > 0$) is *grows exponentially* while anti-Stokes ($\Omega > 0$, $g < 0$) radiation saturates at $\mathcal{S}(L,\omega) = \frac{\hbar\omega}{2\pi} m_{\text{th}}(|\Omega|)$. When losses are taken into account, the gain must overcome a threshold value to enter the amplification regime. Since the Raman gain is frequency dependent, the amplification bandwidth depends on the input power. The effective amplification threshold is usually considered to be reached when Stokes and pump intensity have the same value at the output of the fiber (Agrawal, 2007; Smith, 1972).

By measuring the grows of the Stokes wave, one can deduced the Raman gain as a function of frequency (Mahgerefteh et al., 1996; Stolen et al., 1984). Amplified spontaneous Stokes wave plays an important role in Raman fiber amplifiers (Aoki, 1988; Mochizuki et al., 1986; Olsson & Hegarty, 1986).

Fig. 2 shows the typical (forward) Raman gain $g_\parallel(\omega_p, \Omega)$ and the spontaneous Raman coefficient $R_\parallel(\omega_p, \Omega, T)$ in a silica fiber at $\lambda_p = 1.5$ μm. The parallel Raman gain has a peak at $\Omega_R = 13.2$ THz and a width of about 5 THz. The peak value varies for fiber to fiber. A typical value is $g_R = 1.6$ W^{-1} km^{-1}. The orthogonal gain $g_\perp(\omega_p, \Omega)$ is about 30 times smaller (Agrawal, 2007; Dougherty et al., 1995; Stolen, 1979). The parallel gain can be fit using a 10-Lorentzian model, each Lorentzian having three independent parameters : strength, central frequency, and width (Drummond & Corney, 2001). Note that spontaneous anti-Stokes scattering can be eliminated by lowering the temperature, while the spontaneous Stokes coefficient $R_\parallel$ is at least $\hbar\omega_p/(2\pi) \times g_\parallel$.

2.2 Brillouin scattering

Brillouin scattering is very similar to Raman scattering in the sense it couples two light modes to material vibrations. However, in the contrast with Raman scattering which couples light to molecular vibrations, Brillouin scattering couples light to vibration modes of the fiber itself, that is *sound* waves. Therefore the vibrational frequencies involved in Brillouin scattering are much lower: $|\Omega|/(2\pi)$ is usually in the 10 GHz range. BS is also polarization dependent: as long as the fiber can be considered as mechanically isotropic, their is no orthogonal BS, that is $g_\perp = 0$ (Benedek & Fritsch, 1966; McElhenny et al., 2008; Stolen, 1979). The major difference between Raman and Brillouin scattering lies in the dispersion relation of acoustics vibrations:

$$|\Omega| = v_A |\mathbf{k}_A|, \tag{9}$$

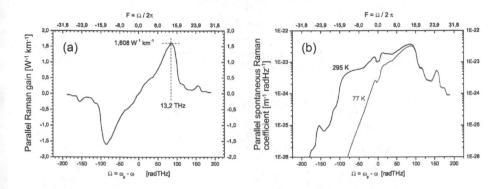

Fig. 2. (a) Raman gain $g_{\parallel}$ in a silica fiber for $\lambda_p = 1.5$ μm and forward propagation. Peak value: $g_R = 1.6$ W^{-1} km^{-1}. Peak position: $\Omega_R = 13.2$ THz. (b) Spontaneous Raman coefficient $R_{\parallel}$ for $\lambda_p = 1.5$ μm and forward propagation at $T = 295$ K and 77 K.

where $v_A = 5.96$ km/s in silica fibers and $\mathbf{k}_A$ is the wave vector of the acoustic wave. Because of momentum conservation, $\mathbf{k}_A$ is equal to the wave vector mismatch between pump and (anti-)Stokes waves: $\mathbf{k}_A = \mathbf{k}_p - \mathbf{k}_{s,a}$. Since the energy difference between pump, Stokes and anti-Stokes waves is very small, $|\mathbf{k}_p| \approx |\mathbf{k}_{s,a}|$ and $|\mathbf{k}_A|^2 \approx 2|\mathbf{k}_p|^2 [1 - \cos(\phi)] = 4|\mathbf{k}_p|^2 \sin^2(\phi/2)$, where ϕ is the angle between $\mathbf{k}_p$ and $\mathbf{k}_{s,a}$. Eq. (9) yields

$$|\Omega| = 2 \, v_A \, |\mathbf{k}_p| \, \sin^2(\phi/2) = 4\pi \, v_A \, \frac{n(\omega_p)}{\lambda_p} \, \sin^2(\phi/2). \tag{10}$$

The maximum value of $|\Omega|$ occurs for backward propagation ($\phi = \pi$), while for forward propagation of (anti-)Stokes waves ($\phi = 0$), $|\Omega| = 0$. Therefore, forward Brillouin scattering is not observed. In the backward direction the Brillouin gain as a peak is at $\Omega_R/(2\pi) = 11.1$ GHz when $\lambda_p = 1.55$ μm. The Brillouin gain has a Lorentzian spectrum

$$g_{\parallel}(\omega_p, \Omega) = \text{sign}(\Omega) \frac{g_B(\Gamma_B/2)^2}{(|\Omega| - \Omega_B)^2 + (\Gamma_B/2)^2} \tag{11}$$

and its spectral width $\Gamma_B/(2\pi)$ is in the 10-100 MHz range. $1/\Gamma_B$ is the decay time of the sound waves. The peak value g_B is usually of the order of 1000 W^{-1} km^{-1}, one thousand times higher than the Raman gain peak g_R.

The Brillouin gain spectrum discussed so far corresponds to a plane acoustic wave propagating along the fiber axis. Other smaller peaks may occur due to other acoustic modes, the presence of dopants and their spatial distribution (Lee et al., 2005; Yeniay et al., 2002). Guided acoustic wave can also produce narrow and very low frequency Brillouin shifts (50 kHz to 1 GHz) and can be even observed in the forward direction (Shelby et al., 1985a;b).

Despite the differences in the Raman and Brillouin gain functions, the underlying scattering mechanism is the same. Therefore, the principle explained in Sec. 3.1 in the context of RS also apply to BS. In particular spontaneous BS exhibits the same temperature dependence as spontaneous RS. Spontaneous Brillouin scattering can be used for temperature sensing (Alahbabi et al., 2005a;b; Pi et al., 2008; Wait et al., 1997).

2.3 Four-photon scattering

The four-photon scattering process differs from the previous scattering processes in that it involves four photons and no material vibration. Since a silica fiber is centro-symmetric, it is the lowest order nonlinear scattering phenomenon that involves only photons in the input and output channels. As shown in Fig. 3, a FPS process consists in the conversion

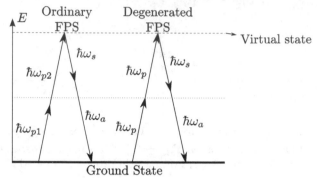

Fig. 3. Spontaneous four-photon scattering processes in silica fibers: ordinary and degenerated case.

of two pump photons at frequencies ω_{p1} and ω_{p2} into two other photons at frequencies ω_s and ω_a. The photon of lower energy is called "Stokes", the one of higher energy in called "anti-Stokes" as in the RS and BS processes. The conversion process satisfies the energy and moment conservation laws

$$\omega_{p1} + \omega_{p2} = \omega_s + \omega_a, \tag{12}$$

$$\mathbf{k}_{p1} + \mathbf{k}_{p2} = \mathbf{k}_s + \mathbf{k}_a. \tag{13}$$

When $\omega_{p1} = \omega_{p2}$ the FPS process is said to *degenerated* and is by far the most studied case, both experimentally and theoretically. FPS is a non resonant process. Therefore many different resonances can contribute to it. In silica, the main contribution comes from electronic resonances. Molecular vibrations contribute to a fraction $f_R = 18\%$ of the FPS strength.

The spectrum of a spontaneous FPS process is usually very broadband. It is not limited by resonance conditions (as RS) or losses (as BS), but merely by the phase matching conditions (13). If a single mode fiber, the wave number of a optical wave has a linear part $k_L(\omega) = n(\omega)\omega/c$ that depends on the effective index $n(\omega)$ of the mode, and a nonlinear part k_{NL} that depends on the power carried by the wave itself (self-phase modulation) and the power of the other waves propagation in the fiber (cross-phase modulation) (Agrawal, 2007). In a spontaneous FPS problem, Stokes and anti-Stokes waves are so faint that their contribution to self or cross-phase modulation is negligible. On the other hand, the pump wave modulates its own phase as well as the phases of the Stokes and anti-Stokes waves. If a wave carries a power P, self-phase modulation changes its own wave number by $k_{NL} = \gamma P$, where γ is the nonlinear coefficient of the fiber. At the same time, that wave modifies the wave number of any other co-polarized wave by $k_{NL} = 2\gamma P$ and any other orthogonally polarized wave by $k_{NL} = (2/3)\gamma P$, through the cross-phase modulation effect. For instance, for a degenerate co-polarized FPS, the wave number mismatch is $\Delta k = k_s + k_a - 2k_p = (k_{Ls} + 2\gamma P_p) + (k_{La} + 2\gamma P_p) - 2(k_{Lp} + \gamma P_p) = \Delta k_L + 2\gamma P_p$, where P_p is the pump power. Using quantum perturbation theory (Brainis, 2009), it can be shown that the spectral density of power at Stokes and anti-Stokes wavelengths is

$$S(L, \omega_s) = S(L, \omega_a) = \frac{\hbar \omega_{s,a}}{2\pi} (\gamma P_p L)^2 \, \text{sinc}^2 \left(\frac{\Delta k}{2} L \right) \tag{14}$$

in the case of degenerate co-polarized FPS. Equivalent formulas for non co-polarized degenerate FPS processes can be found in (Brainis, 2009). Whatever the FPS process (degenerated or not, co-polarized or not, ...), Stokes and anti-Stokes powers are always equal because those photons are created in pairs and the spectrum always depends on the wave number mismatch through the same sinc-function factor, see also Sec. 4.2.

It is important to note that the strength of a spontaneous FPS process scales as $(P_p L)^2$, while the strength of spontaneous RS and BS scales as $P_p L$. The spontaneous FPS spectrum is also independent on temperature. Increasing the propagation length L not only increases the amount of scattered photons, but also narrows the spectrum. In contrast, raising the pump power increases scattering, but as little impact of the spectrum. Therefore, adjusting both parameters, it is possible to set the spectral width of the Stokes and anti-Stokes waves as well as their intensities. Because Stokes and anti-Stokes photons are created in pair, FPS as been extensively studied in the context of photon-pair generation for quantum optics and quantum information applications (Amans et al., 2005; Brainis, 2009; Brainis et al., 2005; 2007; Dyer et al., 2008; Fan & Migdall, 2007; Lee et al., 2006; Li et al., 2004; Lin et al., 2006; 2007; Rarity et al., 2005; Takesue, 2006).

When the scattered intensity becomes high enough ($\gamma P_p \gtrsim 1$), spontaneous scattering gets amplified. In the case of the degenerate co-polarized FPS, the growth of Stokes and anti-Stokes waves in the amplification regime in described by (Brainis, 2009; Dyer et al., 2008)

$$S(L, \omega_s) = S(L, \omega_a) = \frac{\hbar \omega_{s,a}}{2\pi} (\gamma P_p L)^2 \left| \frac{\sinh (g(\omega_{s,a}) L)}{g(\omega_{s,a}) L} \right|^2, \tag{15}$$

where $g(\omega_{s,a}) = \sqrt{(\gamma P_p)^2 - (\Delta k / 2)^2}$ is the *parametric gain* function that appears in the classical theory of four-wave mixing (Agrawal, 2007). Amplification only occurs at those frequencies for which $g(\omega_{s,a}) \in \mathbb{R}$. Because such a condition is never satisfied in the spontaneous regime ($g(\omega_{s,a}) \xrightarrow{P \to 0} i\sqrt{\Delta k_L \gamma P_p}$), it strongly modifies the FPS spectrum when amplification begins. It the amplified regime, the spectral width is determined by $g(\omega_{s,a})$ rather by the propagation length. The peak value of the parametric gain g_P is larger by 70% that the Raman peak gain g_R.

3. Quantum mechanical description of nonlinear scattering

Spontaneous scattering of light cannot be understood in the framework of classical nonlinear optics. A proper description requires the quantum theory. There are two possible approaches. The most elementary one consists in (i) applying quantum perturbation theory to calculate the scattering of light by a single molecule in the first place, then (ii) extending the result to continuous media. The drawback of this method is that it gives access the scattered power density, but not to the field amplitudes. The second approach consists in using a quantum field theory of propagation of light in the fiber that is based on an effective matter/light interaction Hamiltonian.

The "perturbation theory" approach is used in Sec. 3.1 to derive, from first principles, the main formula of Sec. 2.1 for RS. Having identified the limitations of that method, the "field theory method" will be presented in Sec. 3.2.

3.1 Perturbation theory of Raman scattering

The simplest way to model Stokes and anti-Stokes Raman scattering from a coherent pump wave at ω_p consists in applying second order perturbation theory (Crosignani et al., 1980; Wardle, 1999) to matter/light coupling described by the interaction Hamiltonian $H = -\mathbf{d} \cdot \mathbf{E}$, where

$$\mathbf{E}(\mathbf{r},t) = i\frac{f(x,y)}{\sqrt{L}} \left[\sqrt{\frac{\hbar\omega_p}{2\epsilon_0 n^2(\omega_p)}} \, \alpha_p \, e^{i(k(\omega_p)z - \omega_p t)} \, \mathbf{e}_p - c.c. + \sqrt{\frac{\hbar\omega}{2\epsilon_0 n^2(\omega)}} \, a \, e^{i(k(\omega)z - \omega t)} \, \mathbf{e} - h.c. \right]$$

(16)

is the electric field operator associated to the light travelling in the fiber and $\mathbf{d}$ is the electronic dipole moment operator of a scattering fiber molecule at position $\mathbf{r}$. In Eq. (16), the field as been reduced to a pump mode in a coherent state with amplitude α (treated as a strong classical field) and a signal mode representing either the Stokes or anti-Stokes wave at frequency ω. The polarization of the pump and signal modes is defined by the unit vectors $\mathbf{e}_p$ and $\mathbf{e}$. In Eq. (16), the quantity L is the quantization length, a formal parameter that will disappear at the end of the calculation and $f(x,y)$ is normalized so that $\iint f^2(x,y)dA = 1$, where the integration is over the entire fiber cross-section. This normalization is such that

$$\hbar\omega \frac{c}{n(\omega)L} \langle a^\dagger a \rangle = P(\omega) \quad \text{and} \quad \hbar\omega_p \frac{c}{n(\omega_p)L} |\alpha_p|^2 = P_p,$$

(17)

with P_p and $P(\omega)$ the powers in the pump mode and the signal mode, respectively.

The vibration of a molecule can be decomposed in normal modes. Assuming that only one normal mode is excited, the electronic dipole moment of the molecule can be written to first order as

$$\mathbf{d} = \mathbf{d}_0 + \mathbf{d}'Q,$$

(18)

where Q is the normal mode coordinate of the vibration, $\mathbf{d}_0$ the dipole moment around the molecular equilibrium point and $\mathbf{d}' = \frac{\partial \mathbf{d}}{\partial Q}$.

Consider the Stokes process first ($\omega < \omega_p$). Assuming that the molecule starts in the electronic ground state $|g\rangle$ and the vibrational number state $|m\rangle$ and that the Stokes mode is in the Fock state $|n\rangle$, the transition probability amplitude to the state $|g, m+1, n+1\rangle$ after an interaction time t can be calculated using *second* order perturbation theory (Wardle, 1999):

$$c(\Omega, m, n, t) = \frac{f^2(x,y)}{L} \, e^{i[k(\omega)-k(\omega_p)]z} \sqrt{\frac{\omega_p \omega}{8\hbar\epsilon_0^2 n^2(\omega_p)n^2(\omega)M\Omega}} \, \sqrt{m+1} \, \sqrt{n+1} \, \alpha_p$$

$$[\mathbf{e}_p]^\dagger \cdot R \cdot \mathbf{e} \, \frac{e^{i(\Omega+\omega-\omega_p)t} - 1}{\Omega+\omega-\omega_p},$$

(19)

where M and Ω are the effective mass and angular frequency of the molecular normal mode of vibration, while $R \approx 2\sum_e \frac{1}{\omega_{eg}} \langle e|\mathbf{d} \otimes \mathbf{d}' + \mathbf{d}' \otimes \mathbf{d}|g\rangle$, where $\otimes$ denotes the tensor product of two vectors and the sum runs over all the electronic excited states of the scattering molecule having Bohr frequencies ω_{eg} with respect to the ground state.

Around each point $\mathbf{r}$, the material medium is made of many molecules and each molecule has several normal modes of vibration. Since all the molecules are at thermal equilibrium, their vibration have no locked phase relationship. The local field is thus simultaneously coupled to a large thermal reservoir of independent vibration modes and the contribution $c(\Omega, m, n, t)$ of each of them to the overall scattering probability can be added *incoherently*. Writing $\rho(\Omega)$ the number of vibration modes in the volume dV centered on $\mathbf{r}$ with a frequency in the interval $[\Omega, \Omega + d\Omega]$ and and taking the thermal average of the number of excitations in a vibration mode Ω, the *Stokes scattering rate* from the point $\mathbf{r}$ (integrated over all the possible vibration modes frequencies Ω) is found to be

$$S(n;\mathbf{r}) = \frac{dV}{L^2} f^4(x,y) \frac{\pi \omega_p \omega}{4\hbar\epsilon_0^2 n^2(\omega_p)n^2(\omega)} \frac{\rho(|\Omega|)\langle(\mathbf{e}_p^\dagger \cdot R \cdot \mathbf{e})^2\rangle}{M|\Omega|} (m_{\text{th}}(|\Omega|)+1)(n+1)|\alpha_p|^2, \quad (20)$$

where Ω is not an independent variable anymore but is now *defined* as $\Omega := \omega_p - \omega$, and $m_{\text{th}}(|\Omega|)$ – given by Eq. (2) – is the Bose-Einstein expectation value of the number of vibrational excitations. The average $\langle(\mathbf{e}_p^\dagger \cdot R \cdot \mathbf{e})^2\rangle$ is taken over arbitrary molecular orientation in amorphous silica. Therefore the quantity $\langle(\mathbf{e}_p^\dagger \cdot R \cdot \mathbf{e})^2\rangle$ only depends on the angle θ between $\mathbf{e}_p$ and $\mathbf{e}$. As a result, one can write

$$\frac{\rho(|\Omega|)\langle(\mathbf{e}_p^\dagger \cdot R \cdot \mathbf{e})^2\rangle}{M|\Omega|} = K_\parallel(|\Omega|)\cos^2(\theta) + K_\perp(|\Omega|)\sin^2(\theta) = K(|\Omega|,\theta), \quad (21)$$

where $K_\parallel(|\Omega|)$ and $K_\perp(|\Omega|)$ are material characteristics that can be determined experimentally[1]. The total scattering rate from ω_p to ω due to the Stokes process in a fiber segment dz is obtained by integration $S(n, \mathbf{r})$ over the fiber cross-section. As a consequence,

$$S_{p \to s}(n) = dz \frac{\pi \omega_p \omega}{4\hbar\epsilon_0^2 n^2(\omega_p)n^2(\omega)A_{\text{eff}}L^2} K(|\Omega|,\theta)(m_{\text{th}}(|\Omega|)+1)(n+1)|\alpha_p|^2, \quad (22)$$

where $A_{\text{eff}} = 1/\left(\iint f^4(x,y)dA\right)$ is the effective area of the fiber (Agrawal, 2007).

The scattering rate for the anti-Stokes process ($\omega > \omega_p$) can be computed according to the same lines: the rate is the same as in Eq. (22) with the exception that $(m_{\text{th}}(|\Omega|)+1)$ is replaced by $m_{\text{th}}(|\Omega|)$ since an vibrational excitation is destroyed in that process:

$$A_{p \to a}(n) = dz \frac{\pi \omega_p \omega}{4\hbar\epsilon_0^2 n^2(\omega_p)n^2(\omega)A_{\text{eff}}L^2} K(|\Omega|,\theta)m_{\text{th}}(|\Omega|)(n+1)|\alpha_p|^2. \quad (23)$$

During propagation, light is not only scattered from the pump to the Stokes and anti-Stokes modes at $\omega_p - |\Omega|$ and $\omega_p + |\Omega|$ but also *from* these mode to the pump wave. The rates associated to these *reverse* Raman processes are

$$A_{s \to p}(n) = dz \frac{\pi \omega_p \omega}{4\hbar\epsilon_0^2 n^2(\omega_p)n^2(\omega)A_{\text{eff}}L^2} K(|\Omega|,\theta)m_{\text{th}}(\Omega)n|\alpha_p|^2, \quad (24)$$

$$S_{a \to p}(n) = dz \frac{\pi \omega_p \omega}{4\hbar\epsilon_0^2 n^2(\omega_p)n^2(\omega)A_{\text{eff}}L^2} K(|\Omega|,\theta)(m_{\text{th}}(|\Omega|)+1)n|\alpha_p|^2. \quad (25)$$

---

[1] Note that $K_\parallel(|\Omega|)$ and $K_\perp(|\Omega|)$ are slightly different in the core and in the cladding of the fiber because of the dopants. Here, we neglect this difference.

Therefore the *net* Raman scattering rates from a coherent pump to Stokes and anti-Stokes modes (containing n photons initially) are

$$S(n) = S_{p \to s}(n) - A_{s \to p}(n) = \frac{\pi \omega_p \omega dz}{4 \hbar \epsilon_0^2 n^2(\omega_p) n^2(\omega) A_{eff} L^2} K(|\Omega|, \theta) \ (n + m_{th}(|\Omega|) + 1) \ |\alpha_p|^2,$$

(26)

$$A(n) = A_{p \to a}(n) - S_{a \to p}(n) = \frac{\pi \omega_p \omega dz}{4 \hbar \epsilon_0^2 n^2(\omega_p) n^2(\omega) A_{eff} L^2} K(|\Omega|, \theta) \ (-n + m_{th}(|\Omega|)) \ |\alpha_p|^2.$$

(27)

When a pump wave is launched in an optical fiber it scatters photons to many Stokes and anti-Stokes modes simultaneously. The variation in the power spectral density $dS(z, \omega)$ due to the scattering in the fiber slice dz is found by multiplying Eqs. (26) or (27) by the photon energy $\hbar \omega$ and summing over the contribution from all the $c/(n(\omega)L)d\omega$ modes in the interval $[\omega, \omega + d\omega]$. Therefore the following differential equations hold for Stokes and anti-Stokes radiation, respectively:

$$\frac{d}{dz} S(z, \omega) = \begin{cases} \left[S(z, \omega) + \frac{\hbar \omega}{2\pi} (m_{th}(|\Omega|) + 1) \right] g(\omega_p, \Omega, \theta) P_p(z) & \text{if } \Omega = \omega_p - \omega > 0 \text{ (Stokes)} \\ \left[S(z, \omega) - \frac{\hbar \omega}{2\pi} m_{th}(|\Omega|) \right] g(\omega_p, \Omega, \theta) P_p(z) & \text{if } \Omega = \omega_p - \omega < 0 \text{ (anti-Stokes)} \end{cases}$$

(28)

where

$$g(\omega_p, \Omega, \theta) = \text{sign}(\Omega) \frac{\pi}{4 \hbar^2 c^2 \epsilon_0^2} \frac{\omega}{n(\omega_p) n(\omega) A_{eff}} \left(K_{\parallel}(|\Omega|) \cos^2(\theta) + K_{\perp}(|\Omega|) \sin^2(\theta) \right)$$

$$= g_{\parallel}(\omega_p, \Omega) \cos^2(\theta) + g_{\perp}(\omega_p, \Omega) \sin^2(\theta)$$

(29)

is the *Raman gain*. Eq. (28) is identical to Eq. (1).

Unfortunately, BS and FPS laws cannot be established in the same manner. For FPS, one can start the analysis at the molecular scale, but fourth order perturbation theory is required. In addition, transition amplitudes must be added *coherently* to get the phase matching right (see Sec. 2.3). For BS, a molecular approach is not possible since BS couples light to the excitation of an acoustic wave involving many molecules (see Sec. 2.2).

3.2 Nonlinear quantum field theory

The purpose of the quantum field theory approach is to establish a quantum generalization of the nonlinear Schrödinger equation (NLSE) that governs the propagation of the optical field in a fiber, accounting for dispersion and nonlinear interaction with matter, as well as for spontaneous effects.

3.2.1 Operator equation for nonlinear propagation

Kärtner et al. presented a field theory model of Raman scattering (Kärtner et al., 1994). In this model, a light field $A(z, t)$ is coupled to an harmonic field $Q(z)$, the amplitude of which depends on the position in the fiber. The light field $A(z, t)$ represents the envelope of the E-field oscillating at the carrier frequency ω_0 and is assumed to travel in the fiber at group velocity v_g and with no dispersion. Depending on the dispersion relationship of the field $Q(z)$, it can represent acoustical phonons (if $\omega(q) = v_A q$) or optical photons (if $\omega(q) = \Omega_R$

is independent of q). Coupling to acoustical and optical phonons is responsible for BS and RS, respectively. In (Kärtner et al., 1994) it is assumed that $Q(z)$ is an optical phonon field. Physically $Q(z)$ represents the coordinate of a molecular normal mode of vibration at position z in the fiber. Such a field does not propagate but is nevertheless damped. In order to model the damping (with a rate Γ_R), it is assumed that $Q(z)$ itself is coupled to a large bath of harmonic oscillators at many different frequencies that are at thermal equilibrium. These harmonic oscillators represent other optical and acoustical vibration modes. After eliminating the field $Q(z)$ and the bath variables from the equations, one founds that $A(z,t)$ obeys the nonlinear field equation

$$
\frac{\partial}{\partial z} A(z,t) = -\frac{1}{v_g}\frac{\partial}{\partial t} A(z,t) + i(1 - f_R)\gamma A^\dagger(z,t)A(z,t)A(z,t)
$$

$$
+ if_R\gamma \int_{-\infty}^{t} h_R(t-t')A^\dagger(z,t')A(z,t')dt' A(z,t) + i\sqrt{f_R\gamma}N_R(z,t)A(z,t), \tag{30}
$$

and the commutation relationship

$$
\left[A(z,t), A^\dagger(z,t')\right] = \hbar\omega_0\delta(t-t'), \tag{31}
$$

where $f_R = 0.18$ (see Sec. 2.3), γ is the nonlinear coefficient (see Sec. 2.3),

$$
h_R(t) = \frac{\Omega_R^2}{\sqrt{\Omega_R^2 - (\Gamma_R/2)^2}} \sin\left(\sqrt{\Omega_R^2 - (\Gamma_R/2)^2}\, t\right) \exp\left(-(\Gamma_R/2)t\right) v(t) \tag{32}
$$

is the Raman *response function*, and $N_R(z,t)$ is the Raman noise field (see Eqs . (35) and 36 below). In Eq. (32), $v(t)$ is the Heaviside step function. The Fourier transform of $h_R(t)$ is called the Raman susceptibility:

$$
\chi_R^{(3)}(\Omega) = \chi_R'(\Omega) + i\chi_R''(\Omega) = \int_{-\infty}^{\infty} h_R(t)\, e^{-i\Omega t}dt. \tag{33}
$$

Since $h_R(t) \subset \mathbb{R}$ and is normalized such that $\int_{-\infty}^{\infty} h_R(t)dt = 1$,

$$
\chi_R'(-\Omega) = \chi_R'(\Omega), \quad \chi_R''(-\Omega) = -\chi_R''(\Omega), \quad \chi_R'(0) = 1, \quad \chi_R''(0) = 0. \tag{34}
$$

The Raman noise operator is such that its Fourier transform

$$
\tilde{N}_R(z,\Omega) = \int_{-\infty}^{\infty} N_R(z,t)\, e^{-i\Omega t}\, dt \tag{35}
$$

satisfies the following spectral correlations (Boivin et al., 1994; Drummond & Corney, 2001):

$$
\langle \tilde{N}_R^\dagger(z,\Omega)\tilde{N}_R(z',\Omega')\rangle = \hbar(\omega_0 - \Omega)\frac{|\chi_R''(\Omega)|}{\pi}[m_{\text{th}}(|\Omega|) + v(\Omega)]\delta(z-z')\delta(\Omega-\Omega'). \tag{36}
$$

The second term at the right-hand side of Eq. (30) does not come out of Kärtner's model but has been added phenomenologically to account for the $(1 - f_R)$ fraction of the total non linearity that originates in the interaction of light with bound electrons rather than molecular vibrations.

The last term in Eq. (30) is the one responsible for the spontaneous Raman scattering. In order to make the connection with the description given in Secs. 2.1 and 3.1, consider the propagation of a strong continuous pump field $A_p(z,t) = \sqrt{P_p}$ together with a weak scattered field $A_{sc}(z,t)$ that is null at the input of the fiber: $A(z,t) = A_p + A_{sc}(z,t)$. Ignoring all the terms but the last one in Eq. (30) one easily finds that $A_{sc}(z,t) \approx i\sqrt{f_R\gamma P_p}\int_0^L N_R(z,t)dz$. Therefore, the total scattered power is

$$\int_{-\infty}^{\infty} S(L,\omega_0 - \Omega)d\Omega = \langle A_{sc}^\dagger(z,t)A_{sc}(z,t)\rangle = f_R\gamma P_p \int_0^L dz \int_0^L dz' \langle N_R^\dagger(z,t)N_R(z',t')\rangle$$

$$= f_R\gamma P \int_{-\infty}^{\infty} d\Omega \int_{-\infty}^{\infty} d\Omega' \int_0^L dz \int_0^L dz' \langle \tilde{N}_R^\dagger(z,\Omega)\tilde{N}_R(z',\Omega')\rangle$$

Using Eq. (36), one finally gets

$$S(L,\omega = \omega_0 - \Omega) = \hbar(\omega_0 - \Omega)f_R\gamma \frac{|\chi_R''(\Omega)|}{\pi} [m_{th}(|\Omega|) + \nu(\Omega)] P_pL. \tag{37}$$

Comparing this expression with Eq. (5), the Raman gain is found to be related to the imaginary part of the Raman susceptibility by the following relationship:

$$g_\|(\omega_p,\Omega) = -2f_R\gamma\chi_R''(\Omega) \tag{38}$$

According to the Kärtner's model, the Raman gain would be Lorentzian in shape because

$$\chi_R''(\Omega) = \frac{\Omega\Omega_R^2\Gamma_R}{(\Omega_R^2 - \Omega^2)^2 + \Omega^2\Gamma^2}, \tag{39}$$

according to Eqs. (32) and (33). This would be a rough approximation of the actual Raman gain in Fig. 2a. As explained in (Drummond & Corney, 2001), the Raman gain is well fitted by a 10-Lorentzian model. Modifying the quantum field model to couple light to ten Lorentzian vibration modes is trivial: it only changes the shape of the Raman response function $h_R(t)$ in Eq. (32), which becomes a linear superposition of damped sine functions with appropriate oscillation frequencies and damping constants.

With this modification, the quantum propagation equation (30) is able to simulate the spontaneous grow of Stokes and anti-Stokes wave and their amplification. However, Eq. (30) is unable to simulate FPS despite that all the terms (second and third term of the right-hand side) responsible for photon-pair generation are included. This is because phase-matching is of crucial importance for the FPS process and Eq. (30) does not properly deal with the group velocity dispersion of the traveling waves.

3.2.2 Dispersion
Dispersion plays an important role in the physics of spontaneous and stimulated nonlinear effects. The exact dispersion of the fiber can be include in the quantum non linear propagation equation (30) by replacing the first term in the right-hand side

$$-\frac{1}{v_g}\frac{\partial}{\partial t}A(z,t) \tag{40}$$

by the generalized dispersion operator

$$\mathcal{D}[A(z,t)] = +i\sum_{a=1}^{\infty} (i)^a \frac{k_a}{a!}\frac{\partial^a}{\partial t^a}A(z,t), \tag{41}$$

where

$$k_a = \left.\frac{\mathrm{d}^a}{\mathrm{d}\omega^a}k_L(\omega)\right|_{\omega=\omega_0} \tag{42}$$

are the derivatives of the propagation constant $k_L(\omega)$. The dispersion operator can also be written as a convolution integral (Kärtner et al., 1994; Lin et al., 2007)

$$\mathcal{D}[A(z,t)] = \mathrm{i}\int_{-\infty}^{t} h_L(t-t')A(z,t')\mathrm{d}t', \tag{43}$$

where

$$h_L(t) = \frac{1}{2\pi}\int_{-\infty}^{\infty}[k_L(\omega_0 - \Omega) - k_L(\omega_0)]\mathrm{e}^{\mathrm{i}\Omega t}\mathrm{d}\Omega \tag{44}$$

is the *linear* response function of the fiber. Using (43) Eq. (30) reads:

$$\begin{aligned}
\frac{\partial}{\partial z}A(z,t) =& \mathrm{i}\int_{-\infty}^{t}h_L(t-t')A(z,t')\mathrm{d}t' + \mathrm{i}(1-f_R)\gamma A^{\dagger}(z,t)A(z,t)A(z,t) \\
&+ \mathrm{i}f_R\gamma\int_{-\infty}^{t}h_R(t-t')A^{\dagger}(z,t')A(z,t')\mathrm{d}t'A(z,t) + \mathrm{i}\sqrt{f_R\gamma}N_R(z,t)A(z,t),
\end{aligned} \tag{45}$$

3.2.3 Brillouin and polarization effects
As mentioned in Sec. 3.2.1, the quantum propagation model couples light to non propagative phonons. Strictly speaking, such a model is unsuitable for describing BS. However, if the propagation length is long enough to consider that momentum conservation (opto-acoustical phase matching) is verified, the phonon field has a well defined oscillation frequency $\Omega_B = 4\pi\, v_A\,\frac{n(\omega_p)}{\lambda_p}$, see Eq. (10). Therefore, the Brillouin Lorentzian gain can be included as an eleventh Lorentzian (ultra-low frequency) contribution to the nonlinear Raman response $h_R(t)$ (Drummond & Corney, 2001).
Eq. (45) only takes into account nonlinear effects that involve photons with the same polarization state. One can generalize the model to take polarization into account (Brainis, 2009; Brainis et al., 2005; Lin et al., 2006; 2007).

3.2.4 Solving the quantum propagation equation
There are two main methods to solve the quantum nonlinear propagation equation.
The first one is using numerical integration and consists in converting Eq. 45 into a set of c-number equations with *stochastic terms* in order to solve them on a computer (Brainis et al., 2005; Kennedy & Wright, 1988). These methods have been first introduced to solve the scalar quantum equation without the Raman effect ($f_R = 0$) to study the squeezing of a quantum soliton (Carter et al., 1987; Drummond & Carter, 1987) and co-polarized FPS (Brainis et al., 2005). It has been then generalized to study different types of *non* co-polarized FPS processes (Amans et al., 2005; Brainis et al., 2005; Kennedy, 1991) and squeezing in birefringent fibers (Kennedy & Wabnitz, 1988), as well as Raman scattering noise (Drummond & Corney, 2001).
The second method consists in linearizing the quantum nonlinear equation around a classical solution such as a continuous pump wave or a soliton in order to derive linear couple mode operator equations that can be solved analytically (Brainis, 2009; Brainis et al., 2007; Lin et al., 2006; 2007). Coupled mode equations are easier to establish from the Fourier transform of Eq. (45). Defining the Fourier components of the wave as

$$\tilde{A}(z,\Omega) = \int_{-\infty}^{\infty}A(z,t)\mathrm{e}^{-\mathrm{i}\Omega t}\mathrm{d}t, \tag{46}$$

one finds that they satisfy the following equation:

$$\frac{\partial}{\partial z}\tilde{A}(z,\Omega) = i[k_L(\omega_0 - \Omega) - k_L(\omega_0)]\tilde{A}(z,\Omega) + i\sqrt{f_R\gamma}\frac{1}{2\pi}\int_{-\infty}^{\infty}d\omega_1 N_R(z,\Omega - \omega_1)A(z,\omega_1)$$

$$+ i\gamma\frac{1}{(2\pi)^2}\int_{-\infty}^{\infty}d\omega_1\int_{-\infty}^{\infty}d\omega_2\chi(\omega_2 - \omega_1)\tilde{A}(z,\omega_1)\tilde{A}(z,\omega_2)\tilde{A}(z,\Omega + \omega_1 - \omega_2),$$

$$\tag{47}$$

where

$$\chi(\Omega) = (1 - f_R) + f_R\,\chi_R(\Omega) \tag{48}$$

is the total third order susceptibility which takes into account both electronic and vibrational non linearity. $\chi(\Omega)$ is a complex function that has the same symmetry properties as $\chi_R(\Omega)$, see Eq. (34)

$$\chi'(-\Omega) = \chi'(\Omega), \quad \chi''(-\Omega) = -\chi''(\Omega), \quad \chi'(0) = 1, \quad \chi''(0) = 0. \tag{49}$$

Using Eq. (31), one finds that the operators $\tilde{A}(z,\Omega)$ satisfy the following commutation relations

$$\left[\tilde{A}(z,\Omega), \tilde{A}^\dagger(z,\Omega')\right] = 2\pi\hbar\omega_0\delta(\Omega - \Omega'). \tag{50}$$

A generalization of Eqs. (47)-(50) that takes into account polarization can be found in (Lin et al., 2007). Linearized coupled mode equations are directly obtained from Eq. (47). Hereafter, the result is given for one and two pump waves. These coupled-mode equations will be used in Sec. 4 to analyze the competition between the RS process and the FPS process.

3.2.4.1 Single pump configuration

Let us assume that that a monochromatic pump wave with frequency $\omega_p = \omega_0$ and spectral amplitude $\tilde{A}(z = 0,\Omega) = 2\pi\sqrt{P_p}\delta(\Omega)$ is launched in the fiber. During the propagation, the pump remains monochromatic but acquires a nonlinear phase modulation: $\tilde{A}(z,\Omega) = 2\pi A_p(z)\delta(\Omega)$. The amount of phase modulation can be derived by injecting this ansatz in Eq. (47). One finds that

$$\frac{dA_p}{dz} = i\gamma A_p^\dagger(z)A_p(z)A_p(z). \tag{51}$$

The solution of this equation is

$$A_p(z) = \sqrt{P_p}e^{i\gamma P_p z}. \tag{52}$$

However, this solution is not a stable solution of Eq. (47). Brillouin, Raman and four-photon scattering, will spontaneously scattered power from the pump to Stokes and anti-Stokes frequencies. Nevertheless, for the calculation of the Stokes and anti-Stokes amplitudes, one can make the assumption that the pump remains *undepleted*, i.e. (52) is approximately valid. Injecting the ansatz

$$\tilde{A}(z,\Omega) = \left[2\pi A_p(z)\delta(\Omega) + \tilde{A}_{sc}(z,\Omega)\right], \tag{53}$$

and retaining only the terms of highest order in P, one finds that the scattered field $\tilde{A}_{sc}(z,\Omega)$ satisfies

$$\frac{\partial}{\partial z}\tilde{A}_{sc}(z,\Omega) = i\left[[k_L(\omega_p - \Omega) - k_L(\omega_p)] + B(\Omega)\gamma P_p\right]\tilde{A}_{sc}(z,\Omega)$$

$$+ i\chi(\Omega)\gamma P e^{i2\gamma P_p z}\tilde{A}_{sc}^\dagger(z,-\Omega) + i\sqrt{f_R\gamma P_p}e^{i\gamma P_p z}N_R(z,\Omega), \tag{54}$$

where

$$B(\Omega) = \chi(0) + \chi(\Omega) = 1 + \chi(\Omega) = 2 - f_R[1 - \chi_R(\Omega)]. \tag{55}$$

The coefficient $B(\Omega)$ measures the relative strength of the cross-phase modulation of the scattered field by the pump and the self-phase modulation of the pump, see Eq. (52). If Raman scattering is ignored ($f_R = 0$), it takes the usual value $B = 2$. At frequencies close to the pump ($\Omega \to 0$), one also finds $B \approx 2$, because $\chi_R \approx 1 + 0\mathrm{i}$. Very far away from the pump frequency ($\Omega \to \infty$), $B \approx 1.82$ because $\chi_r \approx 0 + 0\mathrm{i}$ and $f_R = 0.18$. The third term of the Eq. (54) represents a FWM process with a complex coupling coefficient $\chi(\Omega)\gamma P_p$. In Sec. 4, it will be shown that this term is responsible for stimulated FWM, stimulated Raman and Brillouin scattering, as well as spontaneous FPS. One may notice that this term couples each spectral component at $\Omega > 0$ (Stokes) to the symmetric component at frequency $\Omega < 0$ (anti-Stokes), as required by the aforementioned processes. The last term in the right-hand side of Eq. (54) is the source of spontaneous Raman and Brillouin scattering. Since Stokes and anti-Stokes frequencies are always coupled, the coupled-mode equations

$$\frac{\partial}{\partial z}\tilde{A}_{sc}(z,\Omega) = \mathrm{i}\,\left[[k_L(\omega_p - \Omega) - k_L(\omega_p)] + B(\Omega)\gamma P_p\right]\,\tilde{A}_{sc}(z,\Omega)$$
$$+\,\mathrm{i}\,\chi(\Omega)\gamma P_p e^{\mathrm{i}2\gamma P_p z}\,\tilde{A}_{sc}^\dagger(z,-\Omega) + \mathrm{i}\,\sqrt{f_R \gamma P_p}e^{\mathrm{i}\gamma P_p z}\,N_R(z,\Omega), \tag{56}$$

$$\frac{\partial}{\partial z}\tilde{A}_{sc}(z,-\Omega) = \mathrm{i}\,\left[[k_L(\omega_p + \Omega) - k_L(\omega_p)] + B(-\Omega)\gamma P_p\right]\,\tilde{A}_{sc}(z,-\Omega)$$
$$+\,\mathrm{i}\,\chi(-\Omega)\gamma P_p e^{\mathrm{i}2\gamma P_p z}\,\tilde{A}_{sc}^\dagger(z,\Omega) + \mathrm{i}\,\sqrt{f_R \gamma P_p}e^{\mathrm{i}\gamma P_p z}\,N_R(z,-\Omega). \tag{57}$$

must be solved together to solve the propagation problem. These are linear, but inhomogeneous equations. Note that

$$\chi(-\Omega) = \chi^*(\Omega), \qquad \text{and} \qquad B(-\Omega) = B^*(\Omega). \tag{58}$$

3.2.4.2 Dual pump configuration

If two pump waves at frequencies $\omega_{p1} = \omega_0 - \Omega_p$ and $\omega_{p2} = \omega_0 + \Omega_p$ are launched simultaneously in the fiber, the spectral amplitude can be written

$$\tilde{A}(z,\Omega) = 2\pi A_{p1}(z)\delta(\Omega - \Omega_p) + 2\pi A_{p2}(z)\delta(\Omega + \Omega_p) \tag{59}$$

Injecting this ansatz in Eq. 47 shows the the two pumps will interact through nonlinear effects:

$$\frac{dA_{p1}}{dz} = \mathrm{i}\,\left[[k_L(\omega_0 - \Omega_p) - k_L(\omega_0)] + \gamma|A_{p1}|^2 + B(2\Omega_p)\gamma|A_{p2}|^2\right]A_{p1}(z) \tag{60}$$

$$\frac{dA_{p2}}{dz} = \mathrm{i}\,\left[[k_L(\omega_0 + \Omega_p) - k_L(\omega_0)] + \gamma|A_{p2}|^2 + B(-2\Omega_p)\gamma|A_{p1}|^2\right]A_{p2}(z) \tag{61}$$

The third terms on the right-hand side are responsible for both cross-phase modulation and stimulated Raman scattering: Eq. (38) shows that

$$\mathrm{i}B(\pm 2\Omega_p)\gamma = \mathrm{i}\,[2 - f_R[1 - \chi'(\pm 2\Omega_p)]]\,\gamma - f_R\gamma\chi''(\pm 2\Omega_p),$$
$$= \mathrm{i}\,[2 - f_R[1 - \chi'(\pm 2\Omega_p)]]\,\gamma + \frac{g_\parallel(\pm 2\Omega_p)}{2}. \tag{62}$$

If the propagation distance L and the initial pump powers P_{p1} and P_{p2} are such that $g_{\parallel}(2\Omega_p)(P_{p1} + P_{p2})L \ll 1$, power transfer due to stimulated Raman scattering is negligible and the solution of Eqs. (60) and (61) is:

$$A_{p1}(z) = P_{p1} \exp\left[i\left[[k_L(\omega_0 - \Omega_p) - k_L(\omega_0)] + \gamma P_{p1} + \Re[B(2\Omega_p)]\gamma P_{p2}\right]z\right], \tag{63}$$

$$A_{p2}(z) = P_{p2} \exp\left[i\left[[k_L(\omega_0 + \Omega_p) - k_L(\omega_0)] + \gamma P_{p2} + \Re[B(-2\Omega_p)]\gamma P_{p1}\right]z\right]. \tag{64}$$

As in the single pump case, such a solution in unstable and light will be spontaneously scattered to other wavelength. To analyze that scattering, we introduce the ansatz

$$\tilde{A}(z,\Omega) = 2\pi A_{p1}(z)\delta(\Omega - \Omega_p) + 2\pi A_{p2}(z)\delta(\Omega + \Omega_p) + \tilde{A}_{sc}(z,\Omega) \tag{65}$$

in Eq. (47) and only keep the terms of highest order in P_{p1} and P_{p2}. It is found that

$$
\begin{aligned}
\frac{\partial}{\partial z}\tilde{A}_{sc}(z,\Omega) = & \; i\left[[k_L(\omega_0 - \Omega) - k_L(\omega_0)] + B(\Omega - \Omega_p)\gamma P_{p1} + B(\Omega + \Omega_p)\gamma P_{p2}\right]\tilde{A}_{sc}(z,\Omega) \\
& + i\gamma\chi(\Omega - \Omega_p)A_{p1}(z)A_{p1}(z)\tilde{A}^{\dagger}_{sc}(z, 2\Omega_p - \Omega) \\
& + i\gamma\chi(\Omega + \Omega_p)A_{p2}(z)A_{p2}(z)\tilde{A}^{\dagger}_{sc}(z, -2\Omega_p - \Omega) \\
& + i\gamma 2\Re\left[\chi(\Omega_p - \Omega)\right]A_{p1}(z)A_{p2}(z)\tilde{A}^{\dagger}_{sc}(z, -\Omega) \\
& + i\gamma\chi(2\Omega_p)A^{\dagger}_{p1}(z)A_{p2}(z)\tilde{A}_{sc}(z, -2\Omega_p) + i\gamma\chi(-2\Omega_p)A^{\dagger}_{p2}(z)A_{p1}(z)\tilde{A}_{sc}(z, 2\Omega_p) \\
& + i\gamma\chi(\Omega_p - \Omega)A^{\dagger}_{p1}(z)A_{p2}(z)\tilde{A}_{sc}(z, \Omega + 2\Omega_p) \\
& + i\gamma\chi(\Omega - \Omega_p)A^{\dagger}_{p2}(z)A_{p1}(z)\tilde{A}_{sc}(z, \Omega - 2\Omega_p) \\
& + i\sqrt{f_R}\gamma\left[A_{p_1}(z)N_R(z, \Omega - \Omega_p) + A_{p_2}(z)N_R(z, \Omega + \Omega_p)\right].
\end{aligned}
\tag{66}
$$

In striking contrast with Eq. (54), the light scattered at frequency $\omega_0 - \Omega$ is not only coupled to the symmetric mode $\omega_0 + \Omega$ but also to six other modes at frequencies: $\omega_0 - 2\Omega_p - \Omega$, $\omega_0 - 2\Omega_p$, $\omega_0 - 2\Omega_p + \Omega$, $\omega_0 + 2\Omega_p - \Omega$, $\omega_0 + 2\Omega_p$, $\omega_0 + 2\Omega_p + \Omega$. As a consequence, there is no way to write down a *closed* set of coupled mode equations for that problem. However, the perturbation theory technique introduced in (Brainis, 2009) can be applied to investigation the quantum regime of scattering (see Sec. 4.2).

4. Coupling between spontaneous scattering processes

When light propagates in an optical fiber, spontaneous RS, BS and FPS take place simultaneously. Several processes may scatter light to the same modes so that it may not be possible the decouple the processes. Hereafter, that point is illustrated in the single and dual pump configuration.

4.1 Single pump configuration
The field evolution in the single pump configuration is fully described by the coupled-mode equations (56) and (57), the solution of which is

$$
\begin{aligned}
\tilde{A}_{sc}(L,\Omega) = & \; \mu_1(L,\Omega)\tilde{A}_{sc}(0,\Omega) + \mu_2(L,\Omega)\tilde{A}^{\dagger}_{sc}(0,-\Omega) \\
& + i\sqrt{f_R\gamma P}\int_0^L N_R(z,\Omega)\left(\mu_1(L-z,\Omega) - \mu_2(L-z,\Omega)\right)dz
\end{aligned}
\tag{67}
$$

$$\tilde{A}_{sc}(L,-\Omega) = \mu_1(L,-\Omega)\tilde{A}_{sc}(0,\Omega) + \mu_2(L,-\Omega)\tilde{A}_{sc}^\dagger(0,\Omega)$$

$$+ i\sqrt{f_R\gamma P} \int_0^L N_R(z,-\Omega)\left(\mu_1(L-z,-\Omega) - \mu_2(L-z,-\Omega)\right)dz. \qquad (68)$$

If $\Omega > 0$, $\tilde{A}_{sc}(L,\Omega)$ corresponds to the Stokes part of the spectrum and $\tilde{A}_{sc}(L,-\Omega)$ to the anti-Stokes part. The functions $\mu_1(z,\Omega)$ and $\mu_2(z,\Omega)$ are defined as

$$\mu_1(z,\Omega) = \left[\cosh(g(\Omega)L) + i\frac{\Delta k(\Omega)}{2g(\Omega)}\sinh(g(\Omega)L)\right]\exp\left(i\frac{k_L(\Omega) - k_L(-\Omega)}{2}L\right) \qquad (69)$$

$$\mu_2(z,\Omega) = i\frac{\gamma\chi(\Omega)P_p e^{i2\gamma P_p L}}{g(\Omega)}\sinh(g(\Omega)L)\exp\left(i\frac{k_L(\Omega) - k_L(-\Omega)}{2}L\right) \qquad (70)$$

where

$$\Delta k(\Omega) = k_L(\omega_p - \Omega) + k_L(\omega_p + \Omega) - 2k_L(\omega_p) + 2\gamma P_p[B(\Omega) - 1] = \Delta k_L(\Omega) + 2\gamma P_p\chi(\Omega) \qquad (71)$$

is the total phase mismatch and

$$g(\Omega) = \sqrt{(\chi(\Omega)\gamma P_p)^2 - (\Delta k(\Omega)/2)^2} = \sqrt{-\Delta k_L(\Omega)\gamma P_p\chi(\Omega) - (\Delta k_L(\Omega)/2)^2} \qquad (72)$$

is the parametric gain. The square-root in Eq. (72) is chosen such that $\Re(g) > 0$. Comparing with the results of Sec. 2.3, both the phase mismatch parameter and the parametric gain have a modified value due to the *simultaneous* action of FPS and RS. This modification impacts the stimulated FWM (Golovchenko et al., 1990; Vanholsbeeck et al., 2003) as well as spontaneous FPS regime (Brainis et al., 2007; Lin et al., 2006).

The spontaneous regime corresponds to the initial conditions $\langle\tilde{A}_{sc}^\dagger(0,\Omega)\tilde{A}_{sc}(0,\Omega)\rangle = 0$, for any value of Ω. The spectral power density $S(L,\omega_p - \Omega)$ at the fiber output can be calculated as follows (Brainis et al., 2005; 2007):

$$S(L,\omega_p - \Omega) = \lim_{\epsilon\to 0}\frac{1}{2\pi\epsilon}\int_{\Omega-\epsilon/2}^{\Omega+\epsilon/2}\int_{\Omega-\epsilon/2}^{\Omega+\epsilon/2}\langle\tilde{A}_{sc}^\dagger(L,\Omega_1)\tilde{A}_{sc}(L,\Omega_2)\rangle d\Omega_1 d\Omega_2. \qquad (73)$$

Using Eqs. (67) and (68), one finds (Brainis et al., 2007)

$$\frac{S(L,\omega_p - \Omega)}{\hbar(\omega_p - \Omega)} = \frac{1}{2\pi}|\chi(\Omega)\gamma P_p L|^2\left|\frac{\sinh(g(\Omega)L)}{g(\Omega)L}\right|^2 + \frac{|\Im[\chi(\Omega)]|\gamma P_p}{\pi}\rho(L,\Omega)\left(m_{th}(|\Omega|) + \nu(\Omega)\right), \qquad (74)$$

where

$$\rho(L,\Omega) = \int_0^L\left|\cosh(g(\Omega)z) + i\,\text{sign}(\Omega)\frac{\Delta k(\Omega)}{2g(\Omega)}\sinh(g(\Omega)z)\right|^2 dz. \qquad (75)$$

The first and second terms in the right-hand side of Eq. (74) represent the photons scattered through the four-photon and Raman processes, respectively. Note that $|\Im[\chi(\Omega)]|\gamma = f_R\gamma|\chi_R''(\Omega)| = |g_\parallel(\omega_p,\Omega)|/2$, see Eq. (38).

In the spontaneous regime ($|g(\Omega)|P_p \to 0$), Eq. (74) reduces to

$$\frac{S(L,\omega_p - \Omega)}{\hbar(\omega_p - \Omega)} = \frac{(|\chi(\Omega)|P_p L)^2}{2\pi}\text{sinc}^2\left(\frac{\Delta k}{2}L\right) + \frac{|g_\parallel(\omega_p,\Omega)|P_p L}{2\pi}\left(m_{th}(|\Omega|) + \nu(\Omega)\right). \qquad (76)$$

The Raman contribution to the scattered light is exactly the one given by Eq. 5: this means that FPS has no impact on RS. The reverse is not true: RS as an influence on FPS since it modifies the total susceptibility $\chi(\Omega)$ appearing in the first term. In absence of RS ($f_R = 0$, $\chi(\Omega) = 1$ and one recovers the spectral density of power given by Eq. (14). Since $-1 < \chi'_R(\Omega) \leq 1$ and $-1.4 < \chi''_R(\Omega) < 1.4$, $|\chi(\Omega)| = \sqrt{(1 - f_R(1 - \chi'_R(\Omega)))^2 + (f_R\chi''_R(\Omega))^2}$ is always close to one. For this reason, spontaneous (not amplified) FPS and RS can be considered has uncoupled phenomena.

If the power is high enough ($|g(\Omega)|P_pL > 1$), amplification of the spontaneous scattering takes place. The general formula (74) is well approximated by

$$\frac{S(L, \omega_p - \Omega)}{\hbar(\omega_p - \Omega)} = \frac{e^{2\Re[g(\Omega)]L}}{8\pi}\left[\left|\frac{\chi(\Omega)\gamma P_p}{g(\Omega)}\right|^2 + \frac{|g_\|(\omega_p, \Omega)|\gamma P_p}{2\Re[g(\Omega)]}\frac{|g(\omega) + i\,\text{sign}(\Omega)\Delta k/2|^2}{|g(\Omega)|^2}\right].$$
(77)

In striking contrast with the analysis of Sec. 2.1, this result shows that the Raman anti-Stokes wave *grows at the same rate as the Raman Stokes wave* instead of saturating at the power density value $\mathcal{S}(L, \omega_p + |\Omega|) = \frac{\hbar\omega}{2\pi}m_{\text{th}}(|\Omega|)$. This effect is due to the coupling of the RS with the FWM. Its detailed explanation can be found in (Brainis et al., 2007; Coen et al., 2002). On the other hand, the exponential amplification of the Stokes wave is completely quenched at frequencies satisfying $\Delta k_L(\Omega) = 0$ because the gain $g(\Omega)$ vanishes in that case, see Eq. (72) (Golovchenko et al., 1990; Vanholsbeeck et al., 2003).

4.2 Dual pump configuration

In the dual pump configuration, the coupled-mode equations (66) do not form a closed set. For this reason, one cannot write an explicit solution as in Sec. 4.1. To study the spontaneous photon scattering, we apply the first-order perturbation technique introduced in (Brainis, 2009).

We first notice that the first term of the right-hand side of (66) represents the phase evolution of scattered field, including the cross-phase modulation due to the two pumps. This phase modulation has no impact on the population of the frequency modes and can be factored out by writing the total scattered E-field

$$E_{\text{sc}}(z, t) = \sqrt{\frac{\hbar\omega_0}{4\pi\epsilon_0 n_0 c}}\int a(z, \Omega)\,e^{i[k_L(\omega_0 - \Omega) + B(\Omega - \Omega_p)\gamma P_{p1} + B(\Omega + \Omega_p)\gamma P_{p2}]z}e^{-i(\omega_0 - \Omega)t}\,d\Omega, \quad (78)$$

where $a(z, \Omega)$ is the annihilation operator of the frequency mode $\omega - \Omega$. Because the exact phase evolution of the $a(z, \Omega)$ has been factored out, the z dependence of $a(z, \Omega)$ is only due to the FPS effect (Brainis, 2009). On the other hand, the scattered field can be written as

$$E_{\text{sc}}(z, t) = \frac{1}{2\pi\sqrt{2\epsilon_0 n_0 c}}\,e^{ik_L(\omega_0)z - i\omega_0 t}\int \tilde{A}_{\text{sc}}(z, \Omega)e^{i\Omega t}\,d\Omega, \quad (79)$$

where we used the fact that $E(z, t) = \sqrt{1/(2\epsilon_0 n_0 c)}A(z, t)e^{ik_L(\omega_0)z - i\omega_0 t}$ and Eq. (46). Comparing Eqs. (78) and (79), one sees that

$$\tilde{A}_{\text{sc}}(z, \Omega) = \sqrt{2\pi\hbar\omega_0}\,a(z, \Omega)\,e^{i[k_L(\omega_0 - \Omega) - k_L(\omega_0) + B(\Omega - \Omega_p)\gamma P_{p1} + B(\Omega + \Omega_p)\gamma P_{p2}]z}. \quad (80)$$

In the following, we make the approximation that $B(\Omega) \approx 2$, see Eq. (55) and $\chi(\Omega) \approx 1$, see Eq. (48). This approximation consists in neglecting the dispersion of the non linearity. Using Eq. (80), one obtains the evolution equation for the annihilation operators:

$$\frac{\partial}{\partial z} a(z, \Omega) = i\gamma \, P_{p1} \, e^{-i\Delta k_{11}(\Omega)z} \, a^{\dagger}(z, 2\Omega_p - \Omega) + i\gamma \, P_{p2} \, e^{-i\Delta k_{22}(\Omega)z} \, a^{\dagger}(z, -2\Omega_p - \Omega)$$

$$+ i\, 2\gamma \sqrt{P_{p1} P_{p2}} \, e^{-i\Delta k_{12}(\Omega)z} \, a^{\dagger}(z, -\Omega)$$

$$+ i\gamma \sqrt{P_{p1} P_{p2}} \, e^{i\Delta k_a z} \, a(z, -2\Omega_p) + i\gamma \sqrt{P_{p1} P_{p2}} \, e^{i\Delta k_b z} \, a(z, 2\Omega_p)$$

$$+ i\gamma \sqrt{P_{p1} P_{p2}} \, e^{i\Delta k_c z} \, a(z, \Omega + 2\Omega_p) + i\gamma \sqrt{P_{p1} P_{p2}} \, e^{i\Delta k_d z} \, a(z, \Omega - 2\Omega_p) \qquad (81)$$

$$+ i\sqrt{\frac{f_R \gamma P_{p1}}{2\pi\hbar\omega_0}} \, e^{i[k(\omega_0 - \Omega_p) - k(\omega_0 - \Omega) - \gamma P_{p1}]z} N_R(z, \Omega - \Omega_p)$$

$$+ i\sqrt{\frac{f_R \gamma P_{p2}}{2\pi\hbar\omega_0}} \, e^{i[k(\omega_0 + \Omega_p) - k(\omega_0 - \Omega) - \gamma P_{p2}]z} N_R(z, \Omega + \Omega_p)$$

where

$$\Delta k_{11}(\Omega) = k_L(\omega_0 - \Omega) + k_L(\omega_0 - 2\Omega_p + \Omega) - 2k_L(\omega_0 - \Omega_p) + 2\gamma P_{p1} \qquad (82)$$

$$\Delta k_{22}(\Omega) = k_L(\omega_0 - \Omega) + k_L(\omega_0 + 2\Omega_p + \Omega) - 2k_L(\omega_0 + \Omega_p) + 2\gamma P_{p2} \qquad (83)$$

$$\Delta k_{12}(\Omega) = k_L(\omega_0 - \Omega) + k_L(\omega_0 + \Omega) - k_L(\omega_0 - \Omega_p) - k_L(\omega_0 + \Omega_p) + \gamma P_{p1} + \gamma P_{p2} \qquad (84)$$

$$\Delta k_a(\Omega) = k_L(\omega_0 + 2\Omega_p) + k_L(\omega_0 + \Omega_p) - k_L(\omega_0 - \Omega_p) - k_L(\omega_0 - \Omega) + \gamma P_{p1} - \gamma P_{p2} \qquad (85)$$

$$\Delta k_b(\Omega) = k_L(\omega_0 - 2\Omega_p) + k_L(\omega_0 - \Omega_p) - k_L(\omega_0 + \Omega_p) - k_L(\omega_0 - \Omega) - \gamma P_{p1} + \gamma P_{p2} \qquad (86)$$

$$\Delta k_c(\Omega) = k_L(\omega_0 - \Omega - 2\Omega_p) + k_L(\omega_0 - \Omega_p) - k_L(\omega_0 + \Omega_p) - k_L(\omega_0 - \Omega) + \gamma P_{p1} - \gamma P_{p2} \qquad (87)$$

$$\Delta k_d(\Omega) = k_L(\omega_0 - \Omega + 2\Omega_p) + k_L(\omega_0 - \Omega_p) - k_L(\omega_0 + \Omega_p) - k_L(\omega_0 - \Omega) - \gamma P_{p1} + \gamma P_{p2} \qquad (88)$$

Eq. (81) can be written

$$i\hbar \frac{\partial}{\partial z} a(z, \Omega) = [G(z), a(z, \Omega)] + i\hbar L(z, \Omega), \qquad (89)$$

where

$$L(z, \Omega) = i\sqrt{\frac{f_R \gamma}{2\pi\hbar\omega_0}} \left[\sqrt{P_{p1}} \, e^{i[k(\omega_0 - \Omega_p) - k(\omega_0 - \Omega) - \gamma P_{p1}]z} N_R(z, \Omega - \Omega_p) \right.$$

$$\left. + \sqrt{P_{p2}} \, e^{i[k(\omega_0 + \Omega_p) - k(\omega_0 - \Omega) - \gamma P_{p2}]z} N_R(z, \Omega + \Omega_p) \right] \qquad (90)$$

and

$$
\begin{aligned}
G(z) =\ & \frac{\hbar}{2}\gamma P_{p1} \int_{-\infty}^{\infty} d\Omega' e^{-i\Delta k_{11}(\Omega')z}\, a^{\dagger}(z,\Omega')a^{\dagger}(z,2\Omega_p - \Omega') + h.c. \\
& + \frac{\hbar}{2}\gamma P_{p2} \int_{-\infty}^{\infty} d\Omega' e^{-i\Delta k_{22}(\Omega')z}\, a^{\dagger}(z,\Omega')a^{\dagger}(z,-2\Omega_p - \Omega') + h.c. \\
& + \hbar\gamma\sqrt{P_{p1}P_{p2}} \int_{-\infty}^{\infty} d\Omega' e^{-i\Delta k_{12}(\Omega')z}\, a^{\dagger}(z,\Omega')a^{\dagger}(z,-\Omega') + h.c. \\
& + \hbar\gamma\sqrt{P_{p1}P_{p2}} \int_{-\infty}^{\infty} d\Omega' e^{i\Delta k_{a}(\Omega')z}\, a^{\dagger}(z,\Omega')a(z,-2\Omega_p) + h.c. \\
& + \hbar\gamma\sqrt{P_{p1}P_{p2}} \int_{-\infty}^{\infty} d\Omega' e^{i\Delta k_{b}(\Omega')z}\, a^{\dagger}(z,\Omega')a(z,2\Omega_p) + h.c. \\
& + \hbar\gamma\sqrt{P_{p1}P_{p2}} \int_{-\infty}^{\infty} d\Omega' e^{i\Delta k_{c}(\Omega')z}\, a^{\dagger}(z,\Omega')a(z,\Omega' - 2\Omega_p) + h.c. \\
& + \hbar\gamma\sqrt{P_{p1}P_{p2}} \int_{-\infty}^{\infty} d\Omega' e^{i\Delta k_{d}(\Omega')z}\, a^{\dagger}(z,\Omega')a(z,\Omega' + 2\Omega_p) + h.c.
\end{aligned}
\tag{91}
$$

In Eq. (89), $L(z,\Omega)$ represents the Raman scattering from both pumps. Raman scattered photons contribute incoherently,

$$
\begin{aligned}
\frac{S_R(L,\omega_p - \Omega)}{\hbar(\omega - \Omega)} =\ & \frac{|g_{\parallel}(\omega_{p1},\Omega - \Omega_p)|P_{p1}L}{2\pi}\left(m_{\text{th}}(|\Omega - \Omega_p|) + v(\Omega - \Omega_p)\right) \\
& + \frac{|g_{\parallel}(\omega_{p2},\Omega + \Omega_p)|P_{p2}L}{2\pi}\left(m_{\text{th}}(|\Omega + \Omega_p|) + v(\Omega + \Omega_p)\right),
\end{aligned}
\tag{92}
$$

to the total scattered photon flux. The $[G(z),a(z,\Omega)]$ part of Eq. (89) represents several FPS processes taking place simultaneously: (i) $2\omega_{p1} \to (\omega_{p1} + \Omega_p - \Omega) + (\omega_{p1} - \Omega_p + \Omega)$, (ii) $2\omega_{p2} \to (\omega_{p2} - \Omega_p - \Omega) + (\omega_{p2} + \Omega_p + \Omega)$, (iii) $\omega_{p1} + \omega_{p2} \to (\omega_0 - \Omega) + (\omega_0 + \Omega)$. To see this explicitly, we writing down the evolution of the quantum state of light in the *interaction* picture. The interaction picture is chosen such that the phase evolution of the modes is part of the operator evolution, while energy transfer from mode to mode is part of the state evolution. In this interaction picture $a^{(I)}(z,\Omega) = a(0,\Omega)$ (Brainis, 2009). Therefore the first order perturbation Dyson expansion gives:

$$
\begin{aligned}
|\psi(L)\rangle =\ & \left(1 + \frac{i}{\hbar}\int_0^L G^{(I)}(z)dz\right)|0\rangle = |0\rangle + \int_{-\infty}^{\infty} d\Omega \\
& \left(\xi_{11}(L,\Omega)|1_{\Omega},1_{2\Omega_p-\Omega}\rangle + \xi_{12}(L,\Omega)|1_{\Omega},1_{-\Omega}\rangle + \xi_{22}(L,\Omega)|1_{\Omega},1_{-2\Omega_p-\Omega}\rangle\right)
\end{aligned}
\tag{93}
$$

where

$$
\xi_{11}(L,\Omega) = i\frac{1}{2}(\gamma P_{p1}L)\, e^{-i\Delta k_{11}(\Omega')\frac{L}{2}}\, \text{sinc}\left(\Delta k_{11}(\Omega')\frac{L}{2}\right)
\tag{94}
$$

$$
\xi_{12}(L,\Omega) = i(\gamma\sqrt{P_{p1}P_{p2}}L)\, e^{-i\Delta k_{12}(\Omega')\frac{L}{2}}\, \text{sinc}\left(\Delta k_{12}(\Omega')\frac{L}{2}\right)
\tag{95}
$$

$$
\xi_{22}(L,\Omega) = i\frac{1}{2}(\gamma P_{p2}L)\, e^{-i\Delta k_{22}(\Omega')\frac{L}{2}}\, \text{sinc}\left(\Delta k_{22}(\Omega')\frac{L}{2}\right).
\tag{96}
$$

The threefold entanglement is a clear signature of the interference between three independent FPS processes. The spectral density of power due to these FPS processes can be deduced from the matrix element $\langle\psi(L)|a^{\dagger}(0,\Omega)a(0,\Omega)|\psi(L)\rangle$ (Brainis, 2009).

5. Conclusion

In this chapter, the physics of Raman, Brillouin, and four-photon scattering processes in silica fibers has been reviewed, as well as their theoretical modeling. It has been shown that a complete quantum field theory is needed to understand the coupling of theses processes in the stimulated and spontaneous regimes. Two examples of coupling have been discussed. The first one was the coupling of the Raman and four-photon scattering processes in a single pump configuration. In that case, it has been shown that the coupling may have spectacular consequences in the amplified spontaneous regime, where an unexpected exponential growth of the anti-Stokes wave is seen. In the second example, the interaction of three FPS processes in a dual pump configuration has been considered. It has been shown that this configuration leads to the generation of a threefold entangled bi-photon state of light.

Spontaneous scattering processes are of great importance the context of quantum light generation and quantum information processing. The methods presented in the chapter apply to the design of quantum source based on optical fibers: engineering the working principle (usually four-photon scattering processes) and estimating their figure of merit (usually limited by the Raman process).

6. Acknowledgement

This research was supported by the Interuniversity Attraction Poles program of the Belgian Science Policy Office, under grant IAP P6-10 "photonics@be" and by the Belgian Fonds de la Recherche Scientifique — FNRS under grant FRFC – 2.4.638.09F.

7. References

Agrawal, G. P. (2007). *Nonlinear Fiber Optics*, fourth edn, Academic Press, Amsterdam.

Alahbabi, M. N., Cho, Y. T. & Newson, T. P. (2005a). 150-km-range distributed temperature sensor based on coherent detection of spontaneous Brillouin backscatter and in-line raman amplification, *J. Opt. Soc. Am. B* 22(6): 1321–1324.

Alahbabi, M. N., Cho, Y. T. & Newson, T. P. (2005b). Simultaneous temperature and strain measurement with combined spontaneous Raman and Brillouin scattering, *Opt. Lett.* 30(11): 1276–1278.

Amans, D., Brainis, E., Haelterman, M., Emplit, P. & Massar, S. (2005). Vector modulation instability induced by vacuum fluctuations in highly birefringent fibers in the anomalous-dispersion regime, *Opt. Lett.* 30(9): 1051–1053.

Aoki, Y. (1988). Properties of fiber Raman amplifiers and their applicability to digital optical communication systems, *J. Ligthwave Technol.* 6(7): 1225–1239.

Benabid, F., Couny, F., Knight, J. C., Birks, T. A. & Russell, P. S. J. (2005). Compact, stable and efficient all-fibre gas cells using hollow-core photonic crystal fibres, *Nature (London)* 434: 488–491.

Benedek, G. B. & Fritsch, K. (1966). Brillouin scattering in cubic crystals, *Phys. Rev.* 149(2): 647–662.

Bloembergen, N. & Shen, Y. R. (1964). Coupling between vibrations and light waves in Raman laser media, *Phys. Rev. Lett.* 12: 504–507.

Boivin, L., Kärtner, F. X. & Haus, H. A. (1994). Analytical solution to the quantum field theory of self-phase modulation with a finite response time, *Phys. Rev. Lett.* 73(2): 240–243.

Bonfrate, G., Pruneri, V., Kazansky, P. G., Tapster, P. & Rarity, J. G. (1999). Parametric fluorescence in periodically poled silica fibers, *Appl. Phys. Lett.* 75(16): 2356–2358.

Brainis, E. (2009). Four-photon scattering in birefringent fibers, *Phys. Rev. A* 79(2): 023840.

Brainis, E., Amans, D. & Massar, S. (2005). Scalar and vector modulation instabilities induced by vacuum fluctuations in fibers: Numerical study, *Phys. Rev. A* 71(2): 23808.

Brainis, E., Clemmen, S. & Massar, S. (2007). Spontaneous growth of Raman Stokes and anti-Stokes waves in fibers, *Opt. Lett.* 32(19): 2819–2821.

Carter, S. J., Drummond, P. D., Reid, M. D. & Shelby, R. M. (1987). Squeezing of quantum solitons, *Phys. Rev. Lett.* 58(18): 1841.

Coen, S., Wardle, D. A. & Harvey, J. D. (2002). Observation of non-phase-matched parametric amplification in resonant nonlinear optics, *Phys. Rev. Lett.* 89(27): 273901.

Corwin, K. L., Newbury, N. R., Dudley, J. M., Coen, S., Diddams, S. A., Weber, K. & Windeler, R. S. (2003). Fundamental noise limitations to supercontinuum generation in microstructure fiber, *Phys. Rev. Lett.* 90(11): 113904.

Crosignani, B., Di Porto, P. & Solimeno, S. (1980). Influence of guiding structures on spontaneous and stimulated emission: Raman scattering in optical fibers, *Phys. Rev. A* 21(2): 594–598.

Dakin, J., Pratt, D., Bibby, G. & Ross, J. (1985). Distributed optical fibre Raman temperature sensor using a semiconductor light source and detector, *Electron. Lett.* 21(13): 569–570.

Digonnet, M. J. F. (ed.) (2001). *Rare-Earth-Doped Fiber Lasers and Amplifiers*, second edn, Marcel Dekker, New-York.

Dougherty, D. J., Kärtner, F. X., Haus, H. A. & Ippen, E. P. (1995). Measurement of the Raman gain spectrum of optical fibers, *Opt. Lett.* 20(1): 31–33.

Drummond, P. D. & Carter, S. J. (1987). Quantum-field theory of squeezing in solitons, *J. Opt. Soc. Am. B* 4(10): 1565.

Drummond, P. D. & Corney, J. F. (2001). Quantum noise in optical fibers. I. Stochastic equations, *J. Opt. Soc. Am. B* 18(2): 139–152.

Dudley, J. M., Genty, G. & Coen, S. (2006). Supercontinuum generation in photonic crystal fiber, *Rev. Mod. Phys.* 78(4): 1135–1184.

Dyer, S. D., Stevens, M. J., Baek, B. & Nam, S. W. (2008). High-efficiency, ultra low-noise all-fiber photon-pair source, *Opt. Express* 16: 9966.

Fan, J. & Migdall, A. (2007). A broadband high spectral brightness fiber-based two-photon source, *Opt. Express* 15: 2915.

Farahani, M. A. & Gogolla, T. (1999). Spontaneous Raman scattering in optical fibers with modulated probe light for distributed temperature Raman remote sensing, *J. Ligthwave Technol.* 17(8): 1379.

Golovchenko, E., Mamyshev, P. V., Pilipetskii, A. N. & Dianov, E. M. (1990). Mutual influence of the parametric effects and stimulated Raman scattering in optical fibers, *IEEE J. Quantum Electron.* 26: 1815–1820.

Hellwarth, R., Cherlow, J. & Yang, T.-T. (1975). Origin and frequency dependence of nonlinear optical susceptibilities of glasses, *Phys. Rev. B* 11(2): 964–967.

Huy, K. P., Nguyen, A. T., Brainis, E., Haelterman, M., Emplit, P., Corbari, C., Canagasabey, A., Ibsen, M., Kazansky, P. G., Deparis, O., Fotiadi, A., Mégret, P. & Massar, S. (2007). Photon pair source based on parametric fluorescence in periodically poled twin-hole silica fiber, *Opt. Express* 15(8): 4419–4426.

Kärtner, F. X., Dougherty, D. J., Haus, H. A. & Ippen, E. P. (1994). Raman noise and soliton squeezing, *J. Opt. Soc. Am. B* 11(7): 1267–1276.

Kazansky, P. G., Russell, P. S. J. & Takebe, H. (1997). Glass fiber poling and applications, *J. Ligthwave Technol.* 15: 1484–1493.

Kennedy, T. A. B. (1991). Quantum theory of cross-phase-modulational instability: Twin-beam correlations in a $\chi^{(3)}$ process, *Phys. Rev. A* 44: 2113–2123.

Kennedy, T. A. B. & Wabnitz, S. (1988). Quantum propagation: Squeezing via modulational polarization instabilities in a birefringent nonlinear medium, *Phys. Rev. A* 38(1): 563.

Kennedy, T. A. B. & Wright, E. M. (1988). Quantization and phase-space methods for slowly varying optical fields in a dispersive nonlinear medium, *Phys. Rev. A* 38(1): 212–221.

Lee, J. H., Tanemura, T., Kikuchi, K., Nagashima, T., Hasegawa, T., Ohara, S. & Sugimoto, N. (2005). Experimental comparison of a Kerr nonlinearity figure of merit including the stimulated Brillouin scattering threshold for state-of-the-art nonlinear optical fibers, *Opt. Lett.* 30(13): 1698–1700.

Lee, K. F., Chen, J., Liang, C., Li, X., Voss, P. L. & Kumar, P. (2006). Generation of high-purity telecom-band entangled photon pairs in dispersion-shifted fiber, *Opt. Lett.* 31: 1905.

Li, X., Chen, J., Voss, P., Sharping, J. & Kumar, P. (2004). All-fiber photon-pair source for quantum communications: Improved generation of correlated photons, *Opt. Express* 12(16): 3737–3744.

Lin, Q., Yaman, F. & Agrawal, G. P. (2006). Photon-pair generation by four-wave mixing in optical fibers, *Opt. Lett.* 31: 1286.

Lin, Q., Yaman, F. & Agrawal, G. P. (2007). Photon-pair generation in optical fibers through four-wave mixing: Role of Raman scattering and pump polarization, *Phys. Rev. A* 75: 023803.

Mahgerefteh, D., Butler, D. L., Goldhar, J., Rosenberg, B. & Burdge, G. L. (1996). Technique for measurement of the Raman gain coefficient in optical fibers, *Opt. Lett.* 21(24): 2026–2028.

McElhenny, J. E., Pattnaik, R. & Toulouse, J. (2008). Polarization dependence of stimulated Brillouin scattering in small-core photonic crystal fibers, *J. Opt. Soc. Am. B* 25(12): 2107–2115.

Mochizuki, K., Edagawa, N. & Iwamoto, Y. (1986). Amplified spontaneous Raman scattering in fiber Raman amplifiers, *J. Ligthwave Technol.* 4(9): 1328–1333.

Olsson, N. & Hegarty, J. (1986). Noise properties of a Raman amplifier, *J. Ligthwave Technol.* 4(4): 396–399.

Pi, Y., Zhang, W., Wang, Y., Huang, Y. & Peng, J. (2008). Temperature dependence of the spontaneous Brillouin scattering spectrum in microstructure fiber with small core, *Tsinghua Science & Technology* 13(1): 43 – 46.

Rarity, J. G., Fulconis, J., Duligall, J., Wadsworth, W. J. & Russel, P. S. J. (2005). Photonic crystal fiber source of correlated photon pairs, *Opt. Express* 13: 534.

Shelby, R. M., Levenson, M. D. & Bayer, P. W. (1985a). Guided acoustic-wave Brillouin scattering, *Phys. Rev. B* 31(8): 5244–5252.

Shelby, R. M., Levenson, M. D. & Bayer, P. W. (1985b). Resolved forward Brillouin scattering in optical fibers, *Phys. Rev. Lett.* 54(9): 939–942.

Smith, R. G. (1972). Optical power handling capacity of low loss optical fibers as determined by stimulated Raman and Brillouin scattering, *Appl. Opt.* 11(11): 2489–2494.

Stolen, R. H. (1979). Polarization effects in fiber Raman and Brillouin lasers, *IEEE J. Quantum Electron.* 15: 1157–1160.

Stolen, R. H., Lee, C. & Jain, R. K. (1984). Development of the stimulated Raman spectrum in single-mode silica fibers, *J. Opt. Soc. Am. B* 1(4): 652–657.

Stolen, R. & Ippen, E. (1973). Raman gain in glass optical waveguides, *Appl. Phys. Lett.* 22(6): 276–278.

Takesue, H. (2006). Long-distance distribution of time-bin entanglement generated in a cooled fiber, *Opt. Express* 14: 3453.

Vanholsbeeck, F., Emplit, P. & Coen, S. (2003). Complete experimental characterization of the influence of parametric four-wave mixing on stimulated Raman gain, *Opt. Lett.* 28: 1960–1962.

Wait, P. C., Souza, K. D. & Newson, T. P. (1997). A theoretical comparison of spontaneous Raman and Brillouin based fibre optic distributed temperature sensors, *Opt. Commun.* 144(1-3): 17 – 23.

Wardle, D. A. (1999). *Raman Scattering in Optical Fibres*, PhD thesis, University of Auckland.

Yeniay, A., Delavaux, J.-M. & Toulouse, J. (2002). Spontaneous and stimulated Brillouin scattering gain spectra in optical fibers, *J. Lightwave Technol.* 20(8): 1425.

4

Progress in Continuous-Wave Supercontinuum Generation

Arnaud Mussot and Alexandre Kudlinski
Laboratoire PhLAM, IRCICA, Université Lille 1
France

1. Introduction

Supercontinuum (SC) light sources are nowadays a very common way to access a large span of wavelengths, usually ranging from the near ultraviolet (around 400 nm) to the infrared (around 2.4 μm). It corresponds to a range of interest for many applications in optics for measuring transmission, dispersion, or in biophotonics for achieving fluorescence microscopy, optical coherence tomography... Indeed these sources are really promising because they should allow to replace the N laser sources used in these experimental setups to access to all theses wavelengths by a single broad one and a spectral filtering apparatus. Most of these results have been obtained by using powerful pump lasers of several kilo-Watts peak power, operating from the femtosecond (Titane:Saphire) to the nanosecond regimes (Nd:YAG), launched in the low dispersion region of a microstructured optical fiber. Although these fibers are short enough (typically from 1 to 10 m) to neglect the linear absorption during the propagation of the pump, the spectral power density is relatively low (few hundreds of μw/nm) which could limit the implementation of SC sources in many application devices. This is related to a technological limitation of the pump source because it is not easy to combine strong peak power and high average power. One of the simplest solutions to increase the the spectral power density of SC sources is to replace pulsed sources with continuous-wave (CW) light sources whose available average powers are much more important. We will see that the dynamics of SC formation is considerably different in this case, requiring to perform intensive numerical studies to optimize the fiber parameters. Indeed, longer fibers are required (from tens to hundreds of meters) which heightens sensitivity to fiber attenuation, namely of the OH pic absorption, that strongly impacts the soliton evolution. However extremely powerful SCs have been reported with more than 10 mW/nm of spectral power density. Furthermore, these pump sources are usually all-fiber that leads to a second advantage against most of pulsed SC because CW pump can be directly spliced on the PCF. It is also important to point out that these SC sources have different temporal properties than the ones of pulsed SCs.

The first experimental demonstration of CW SC have been realized at the end of the nineties with a Raman laser launched in a standard telecommunication fiber. The spectral broadening was relatively restricted (around 200 nm) because it was mainly due to Raman effect Gonzalez-Herraez et al. (2003); Persephonis et al. (1996); Prabhu et al. (2000). A breakthrough was reached a few years later when stronger pump lasers (from more than one order of magnitude) based on Ytterbium doped fibers were combined with photonic crystal fibers

(PCFs) owing a low group-velocity dispersion (GVD) value around the pump wavelength Avdokhin et al. (2003). With these setups, SC generation was mainly due to solitonic effects like in pulsed SC. A renew of interest for these sources started from 2007 where first numerical demonstrations of very broad CW SC were reportedMussot et al. (2007), just followed by experimental demonstrationsCumberland et al. (2008a); Kudlinski & Mussot (2008).

As an example, a typical and simple experimental setup used for SC generation is schematized in Fig. 1. In our experimental configuration used for the experiments hereafter, the PCFs were

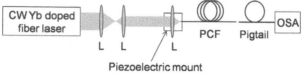

Fig. 1. Scheme of the experimental setup used for the SC generation experiments. L : lens.

pumped with ytterbium-doped fiber laser delivering either 20 W at 1064 nm with a full width at half maximum of 0.5 nm, or 50 W or 100 W at 1070 nm with a full width at half maximum of 1 nm. The output beam diameter of the laser was reduced with an afocal setup and the beam was then launched into the fiber with appropriate aspheric lenses (of a few mm focal length). All lenses were antireflection coated and a heat dissipater was carefully placed on top of the V-groove supporting the PCF in order to manage thermal issues caused by the high power laser. This allowed to greatly improve the temporal stability of the injection setup and no noticeable change in coupling efficiency was observed for several tens of minutes at full pump power. The coupling efficiency in these conditions was typically 70%–80%. The output of the fiber was butt-coupled to a pigtail to reduce the power launched inside the optical spectrum analyzer (OSA). All-fiber schemes are also used Cumberland et al. (2008a) but splicing issues are usually more -time-consuming for a lab experiment than free-space coupling.

2. Basic mechanisms of continuous-wave supercontinuum generation

Mechanisms at the origin of SC are now well knownCumberland et al. (2008b); Dudley et al. (2006); Kobtsev & Smirnov (2005); Mussot et al. (2007); Travers et al. (2008); Vanholsbeeck et al. (2005). The modulationnal instability (MI) process is at the origin of the formation of CW SC. It originates from the perfect balance between linear and nonlinear effects experienced by a strong field, the pump, and a small perturbation, the noise, when working in anomalous GVD region of an optical fiber. At the beginning of the fiber, this small perturbation is amplified. The typical signature of this process in the spectral domain is two symmetric side lobes located around the pump (Figs. 2-(a) and (b)). By further propagating into the fiber, this small periodic perturbation is amplified to become a train of solitonic pulses (Figs. 2-(c)and (d)). Note that these pulses have not identical characteristics as they originate from a process that is seeded by noise. On the other hand, we remind that in the case of a single soliton propagating in an optical fiber, it is well known that during its propagation, it is disturbed by higher order dispersion orders effect and as a consequence it shed energy to radiations called dispersive waves (DWs) which verify a phase matching condition. This leads to the generation of DWs on the short wavelength side of the soliton and to a red shift of the soliton, called *spectral recoil* for momentum conservation (Figs. 3-(a) and (b)). The consequence of the phase matching condition, is that solitons and DWs do not travel at the same velocity. If no additional effect is experienced by these waves, they will no longer interact again during their respective

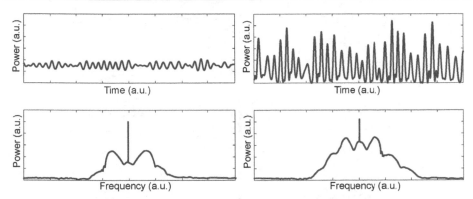

Fig. 2. Schemes illustrating the first steps of modulationnal instability, (a)-(b) at the beginning of the process when the small perturbation starts to grow and (c)-(d) when the solitonic train is created. (a) and (c) correspond to the time domain, and (b) and (d) correspond to the frequency domain.

propagation. However, during the propagation of solitons, because they are strong waves, the Raman effect induces an additional and continuous red shift of their central frequencies . This effect is called soliton self-frequency shift (SSFS)Gordon (1986) and decelerates the soliton because the group index increases with frequency in usual fibers (Fig. 3-(c)). As a consequence, as the velocity of DWs has not changed, the solitons can interact again with them via the cross-phase modulation effectGenty et al. (2004) which shifts DWs toward short wavelengths. Finally, it is important to understand that this group velocity matching is the rule that allows to connect lower and upper limits of SCs (Figs. 3-(d))Stone & Knight (2008). This is different in fibers with two zero dispersion wavelengths (ZDWs). When the soliton approaches the second ZDW, it still generates DWs but on the long wavelength side of the SCGenty et al. (2004); Mussot et al. (2007). The spectral recoil tends now to shift it on the opposite side that the one of the SSFS. An equilibrium is reached and the frequency shift of the soliton is cancelledSkryabin et al. (2003). In this case solitons and DWs will no interact together and no trapping mechanism will occur like it is the case in fiber with a single ZDW.

3. Bandwidth-limited near infrared continuous-wave supercontinuum

The interest of limiting the spectral extension of CW SC is to concentrate of the available power in the desired spectral span. This is achieved in fibers with two ZDWs in which the SSFS can be cancelled by the spectral recoil effect experienced by solitons located just below the second zero dispersion wavelength. By this way, increasing the power leads to an increase of the power spectral density.

3.1 Double zero-dispersion wavelength photonic crystal fibers

It is possible to design PCFs with two ZDWs from part to part of the pump wavelength at 1064 nm, and a low anomalous dispersion region at this wavelength. Such group-velocity dispersion (GVD) curves can be achieved with a relatively small hole-to-hole spacing Λ in the order of 1.7 μm and a d/Λ value in the range of 0.4–0.5 (d is the hole diameter) Mussot et al. (2007); Tse et al. (2006). It is well known that a microstructured cladding with these geometrical properties would lead to relatively high confinement losses at wavelengths around 1.5 μm which is the reason why 10 periods of holes were necessary between the core

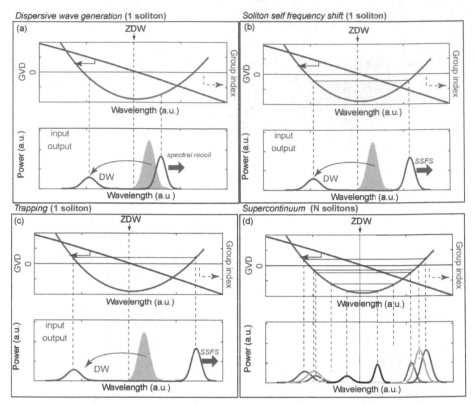

Fig. 3. Schemes illustrating (a) the DW generation from a single soliton, (b) the SSFS effect and (c) the trapping of teh DW by the soliton, *for a single soliton*. In each cases, the dispersion curve is represented in blue and the group index curve in red. (d) Schemes illustrating the formation of a CW SC, *involving N solitons*.

and the external jacket to decrease confinement losses to an acceptable values of a few dB/km at 1550 nm, at which the SC is expected to be generated. Figure 4(a) displays a scanning electron microscope (SEM) image of such a PCF (labelled fiber C in what follows, see Table 1). Another important issue in CW SC generation is the absorption of the water band centered at 1380 nm Cumberland et al. (2008a). We thus performed a chemical cleaning of the stacked preform under halogenic atmosphere to reduce surface contamination and to lower the water content. This allowed to decrease the peak attenuation at 1380 nm from typically 600 dB/km (without any special treatment) to about 120 dB/km. A typical attenuation spectrum is shown in Fig. 4(b) for fiber C, which SEM is represented in Fig. 4(a). The background loss being around 30 dB/km at 1380 nm, the contribution of the water contamination is about 90 dB/km at 1380 nm.

Three different PCF samples (labeled A, B and C) are investigated here. These fibers are characterized by slightly differing Λ and d/Λ values so that the GVD curve of each fiber is slightly different. The GVD curves are represented in Fig. 5(a), where the vertical line represents the pump wavelength. All GVD curves have been calculated with a finite-elements

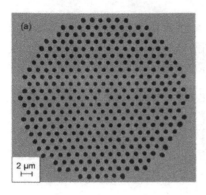 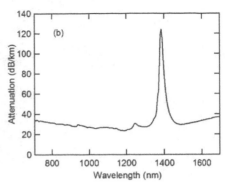

Fig. 4. (a) Typical SEM image of PCFs used for infrared CW supercontinuum. (b) Typical attenuation spectrum.

Parameter	Fiber A	Fiber B	Fiber C
ZDW1 (nm)	1012	958	903
ZDW2 (nm)	1236	1346	1570
γ at 1064 nm ($W^{-1}.km^{-1}$)	22	24	30
GVD at 1064 nm (ps/nm/km)	+4	+12	+30

Table 1. Parameters of the PCFs under investigation in this section.

method (FEM) from high resolution SEM images of the PCFs. Table 1 summarizes the properties of the three fibers under investigation.

3.2 Control of the supercontinuum long-wavelength edge

The experiments reported in this section were performed in 100 m-long samples of each PCF described in Table 1. In the launching conditions described in the previous paragraph, the output power were respectively 7.35 W, 7.25 W and 7.16 W for fibers A to C, with a 20 W CW fiber laser at 1064 nm. Figure 5(b) shows the spectra obtained in all fibers. The green curve corresponds to fiber A, with the closest ZDWs. As expected from Ref. Mussot et al. (2007), the spectral width is limited by the second ZDW. The dispersion value at the 1064 nm pump wavelength is very low (+4 ps/nm/km). The spectral broadening is thus initially dominated by MI, with the anti-Stokes MI sideband overlapping with the normal GVD region. The short wavelength extension just below the first ZDW results from a spectral overlap of MI sidebands and blue-shifted dispersive wave in the normal GVD region, as analyzed in Cumberland et al. (2008b). Since the spectral position of the Stokes MI sidebands is just below the second ZDW of the fiber, the SSFS is very short and the spectrum remains consequently quite symmetric. The power generated above the second ZDW (depicted by a vertical line) is attributed to the generation of red-shifted dispersive waves accompanying the cancelation of the SSFS. The red curve in Fig. 5(b) corresponds to the spectrum measured for fiber B, with both ZDWs separated by about 400 nm. In this fiber, the long-wavelength ZDW is very close to the center of the water absorption band (1380 nm). It is well known that the SSFS is canceled if a second ZDW is present at longer wavelengths Skryabin et al. (2003), which is the case here. This is seen in the red spectrum of Fig. 5(b) as a peak centered at 1310 nm, which corresponds

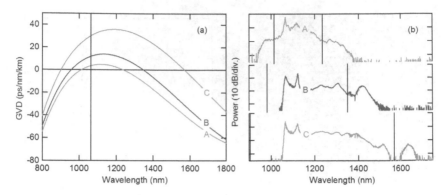

Fig. 5. (a) GVD curves of the three PCFs under investigation in this section. The vertical line depicts the pump wavelength. (b) Corresponding output spectra recorded for a fiber length of 100 m and a pump power of 12.5 W launched into the PCFs. Vertical lines represent the ZDWs.

to an accumulation of all solitons stopped by the second ZDW. Additionally, a large part of energy is transferred to a dispersive wave centered at 1420 nm, which is phase-matched with the solitons. In fiber C displayed in Fig. 5(b) in blue, the flatness of the spectrum is clearly affected by the water absorption at 1380 nm. This can be seen as a spectral decrease at wavelengths higher than 1380 nm. The higher energy solitons are able to tunnel through the water attenuation band and are then stopped by the second ZDW located at 1570 nm in fiber C. The less energetic ones stop just below 1380 nm due to the water absorption peak. As in fiber B, an accumulation of solitons is seen just before the second ZDW, and a dispersive wave which is phase-matched with the solitons is generated at 1630 nm. Note that the peak located around 1120 nm is due to Raman lasing because of Fresnel reflection at both fiber faces. It is also important to note that, unlike in fiber A, no short wavelength extension is observed in fibers B and C. Indeed, in these fibers, the amount of energy transferred from solitons to blue-shifted dispersive waves is negligible because there is no spectral overlap between solitons and dispersive waves Akhmediev & Karlsson (1995).

3.3 Dynamics of the supercontinuum formation

To go further into the detailed dynamics of SC formation, we performed a cut-back measurement on fiber C. The spectrum was measured every 5 m for fiber lengths between 0.5 and 100 m. The results are displayed in Fig. 6, where the output spectra are represented as a function of fiber length. The solitonic waves created by MI are progressively red-shifted by SSFS during the first 30 m of propagation. They are then stopped by the second ZDW located at 1570 nm (depicted by the white dotted vertical line). The soliton build-up due to spectral recoil before the second ZDW can be seen as an increase in spectral power. The red-shifted dispersive wave is also observed from this propagation length of 30 m. For more important fiber lengths, the spectrum extension remain almost constant. The experimental results displayed in Figs. 5 and 6 illustrate the possibility of tailoring the spectrum extent in the context of multi-watt and relatively flat SC generation. The long-wavelength edge of the spectrum is limited by red-shifted dispersive waves, whose spectral location is imposed by the second ZDW.

4. Extension towards visible wavelengths

Another long term issue of CW-pumped SC concerns the lack of short wavelengths generation when pumping at 1 μm. The generation of visible wavelengths would be of great interest for a substantial number of applications including high resolution imaging, metrology or spectroscopy. One possible approach to achieve this is to take advantage of the process of dispersive wave trapping by solitons Nishizawa & Goto (2002). This process leads to an extra blue shift of dispersive waves in the spectral domain Genty et al. (2004; 2005); Gorbach & Skryabin (2007a;b); Gorbach et al. (2006); Travers (2009); Travers & Taylor (2009). Experimentally, this phenomenon has been proved to be of primary importance to generate short wavelengths in pulsed pumping regime Stone & Knight (2008). It has also been combined with dispersion-engineered PCFs to further extend SC to the UV in nanosecond and picosecond pumping schemes Kudlinski et al. (2006). This idea consists in modifying the dispersion curve along the fiber so that group-velocity matching conditions for trapped dispersive waves continuously evolve along propagation. This leads to the generation of new wavelengths as the ZDW decreases along propagation. The present work is based on this idea which has been adapted to CW pumping conditions.

4.1 Zero-dispersion wavelength decreasing photonic crystal fibers

The dispersion-engineered PCF firstly used within this framework consists of a 100 m-long section with a constant dispersion followed by a 100 m-long section with decreasing ZDW, as illustrated in Fig.7(c). The total attenuation of the 200 m-long PCF is 1.5 dB at 1064 nm. A SEM image of the input and output faces of the PCF is represented in Figs. 7(a) and (b) respectively, with the same scale. The input outer diameter is 125 μm, the hole-to-hole spacing Λ is 4.7 μm and the hole diameter d is 2.6 μm. The dispersion curves at the PCF input and output have been computed with a finite elements method from high resolution SEMs and are represented

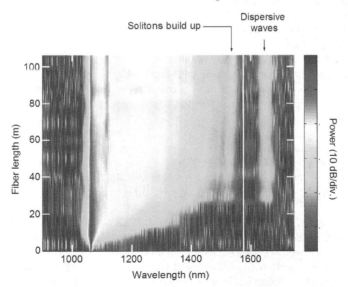

Fig. 6. Experimental measurement of the SC dynamics as a function of fiber length, in fiber C. The white vertical line represent the second ZDW.

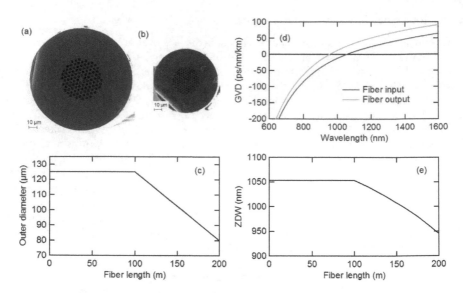

Fig. 7. (a),(b) SEMs of the input and output faces of the ZDW decreasing fiber. Respective outer diameters are 125 and 80 μm. (c) Outer diameter versus fiber length. (d) GVD curves at the input (red line) and output (blue line) of the ZDW decreasing fiber. (e) ZDW versus fiber length.

in Fig. 7(d) in red and blue lines, respectively. The input ZDW is located at 1053 nm, just below the pump wavelength of 1064 nm. To decrease the size of the microstructure along propagation and consequently shift the ZDW toward shorter wavelengths, the outer diameter of the fiber has been approximately linearly reduced to a final diameter of 80 μm (see Fig. 7(c)). This was done by gradually increasing the drawing speed during the fiber fabrication whilst keeping the preform feed rate constant. The pitch Λ at the fiber output was reduced to 3.1 μm and the d/Λ ratio was kept constant along the whole PCF, so that the output ZDW is shifted down to 950 nm. The longitudinal evolution of the ZDW of the PCF is represented in Fig. 7(e). In the first 100 m, the ZDW is fixed to 1053 nm, and it drops to 950 nm in a quasi-linear way along the last 100 m.

4.2 Generation of visible light
The setup used to pump the fabricated PCF is shown in Fig. 1. The beam from a 20 W CW fiber laser at 1064 nm was collimated and launched into the fiber with a lens of 4.5 mm focal length. The coupling efficiency was 75%, corresponding to a power of 13.5 W launched into the PCFs.

The SC spectrum measured at full pump power in the ZDW-decreasing PCF described above displayed in Fig. 8 in red line. It is ranging from 670 nm to 1350 nm with an output power of 9.5 W. Additional spectral components are located around 550 nm and the visible part of the SC was easily observable with naked eye at the PCFs output. The inset of Fig 8 displays a far-field image of the whole visible spot observable at the PCF output. Since the modal distribution of the SC does not look single mode, the far-field profile was investigated as a function of wavelength by using 10 nm bandpass filters. Right insets of Fig. 8 show far-field

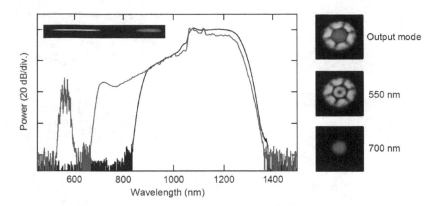

Fig. 8. Output spectra obtained in the uniform PCF (black line) and in the ZDW-decreasing one (red line). Inset: output beam dispersed by a prism. Right: output far-field without any filter (top), with a 550 nm filter (center) and with a 700 nm filter (bottom).

images centered at 550 nm and 700 nm . The green spectral components located around 550 nm are clearly generated in higher-order modes. This part of the spectrum was not expected from our design and is probably due to a phase-matching condition satisfied with higher-order modes Efimov et al. (2003); Omenetto et al. (2001). The far-field image recorded at 700 nm shows that the red spectral components are generated in a fundamental mode. We also checked that the mode was fundamental-like in the whole spectrum above 700 nm for all fibers.

A uniform PCF with dispersion comparable to the ZDW-decreasing fiber input was used for comparison. The output spectrum obtained in the same conditions than above is displayed in Fig. 8, in grey line. It extends up to 1355 nm into the infrared, which is very similar to the spectrum obtained in the ZDW-decreasing PCF. However, the short-wavelength edge is located at about 840 nm, which is much less spectacular than in in the ZDW-decreasing PCF where it reaches 670 nm. This clearly shows that the extra 170 nm bandwidth toward the visible is generated thanks to the decreasing ZDW.

In order to generate even shorter wavelengths, the ZDW-decreasing PCF has been pumped with a more powerful Yb fiber laser delivering 50 W at 1070 nm with a full width at half maximum of 1 nm. With the same setup as described above, we were able to launch a maximum power of 35 W in the fiber, corresponding to a coupling efficiency of 70 %. The resulting experimental spectrum is displayed in Fig. 9. For the highest pump power of 35 W, the SC is ranging from 650 nm to 1380 nm with a 19.5 W output power.

4.3 Discussion and numerical modelling

As claimed above, the basics of using a ZDW decreasing PCF to extend the spectrum towards short wavelengths was to use progressively red-shifted solitons to trap dispersive waves in the visible. The results displayed in Fig. 8 indeed suggest that the long- and short-wavelength edges of the spectra are correlated. In order to have a further insight into the mechanisms of the visible SC formation, the power dynamics of the spectral broadening has been investigated in the ZDW-decreasing PCF for launched powers of 8.2 W, 11.3 W and 13.5 W (see Ref. ?). As expected, a broadening of the output spectrum occurs on both sides with increasing

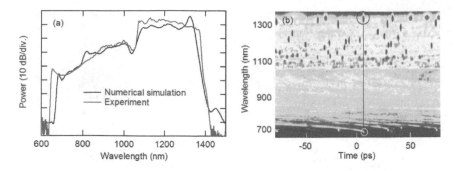

Fig. 9. Experimental (red line) and numerical (black line) output spectra obtained in the ZDW-decreasing PCF for a launched pump power of 35 W. The average power at the fiber output is 19.5 W. Spectrogram of a single-shot simulation at the output of the 200 m-long ZDW-decreasing PCF. The vertical line joins a soliton and its corresponding trapped dispersive wave. The color scale ranges over 30 dB.

launch power inside the fiber. The long and short wavelengths sides were identified by measuring the wavelength of typical spectral features on both edges of the spectrum. The spectral broadening on the long-wavelength side stops at 1140, 1160 and 1250 nm for respective increasing pump powers. The short-wavelength edge is progressively blue-shifted with increasing pump powers and extends to respectively 763, 751 and 720 nm. These experimental results have been compared with the computed group-index curve calculated for the end (small diameter) of the ZDW-decreasing PCF. This is illustrated in Fig. 10, where the group-index curve is plotted as a function of wavelength. Markers represents the long and short wavelength edges experimentally measured for launched powers of 8.2 W (green), 11.3 W (blue) and 13.5 W (red) respectively. The corresponding points for a fixed power are joined by nearly horizontal lines on the plot, which means that these radiations travel at almost the same group velocity when they go out of the fiber. This provides a strong support to the process of group-velocity matching between the most blue- and red-shifted spectral components of each spectrum and this evidences the benefit of the ZDW-decreasing fiber for the generation of shorter wavelengths.

In order to get a deeper understanding of the nonlinear mechanisms originating the visible extension, we performed numerical simulations. We integrated the generalized nonlinear Schrödinger equation including the experimental attenuation curve and all experimental parameters. The numerical method used to model the pump laser was fully described in Ref. Mussot et al. (2007). The simulations were very time consuming (1 week on a standard PC) so we did not perform the usual averaging procedure Mussot et al. (2007) required to account for the experimental measurements. The output spectra resulting from a single simulation is plotted in black line in Fig. 9(a). By performing several other simulations we checked that there was no significant modification from simulation to simulation. The agreement between numerical simulations and experimental results is excellent in terms of shape and extension of the spectrum. It is important to note that numerical simulations have been performed without any free parameters. In both cases, the stop of the spectral broadening at the long-wavelength is due to a relatively high OH absorption around 1380 nm (measured to be about 300 dB/km). The slight discrepancy with experiments observed at short wavelengths (680 nm for the simulations against 650 nm for the experiments) is

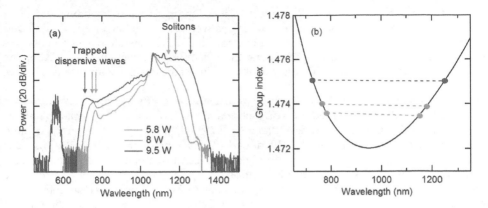

Fig. 10. (a) Output spectra recorded in the ZDW-decreasing PCF for pump powers of 8.2 W (green), 11.3 W (blue) and 13.5 W (red). Output powers are indicated on the graph. (b) Group-index curve calculated at the ZDW-decreasing PCF output (black line). Markers indicate extreme wavelengths of the corresponding SC spectrum experimentally recorded.

probably due to the uncertainty of the calculated GVD curve, leading to a slightly different group-velocity matching condition between red-shifted solitons and trapped dispersive waves. To further illustrate the trapping mechanism responsible for the short wavelength part of the spectrum, the numerical spectrogram of the optical field at the output of the fiber is represented in Fig. 9(b). It corresponds to the numerical spectrum displayed in black line in Fig. 9(a). In the spectrogram, one can see a whole spectral region full of solitons (represented as red dots) originating from the initial MI process. This region extends from the pump wavelength (1070 nm) to the upper limit of the spectrum. Some of the solitons are close to the pump wavelength and some other ones exhibit an important red-shift due to SSFS. The most shifted ones are stopped by the important OH absorption peak at 1380 nm. The region between 680 nm and the pump wavelength corresponds to dispersive waves generated from solitons Travers (2009); Travers et al. (2008). For the most red-shifted solitons, a blue-shifted trapped dispersive wave can be observed just below 700 nm, both travelling at the same group-velocity. An example of a soliton group-velocity matched with a dispersive wave is highlighted in Fig. ??, where both are joined by a black line. We can see that the trapped dispersive wave is exactly at the vertical of the soliton which confirms that both waves travel at the same velocity inside the fiber. It thus confirms that the extension of the SC towards short wavelengths is mainly due to the trapping of dispersive waves by red-shifted solitons Nishizawa & Goto (2002) rather than by the basic dispersive wave generation process Akhmediev & Karlsson (1995). It should be noted then that the generation of even shorter wavelengths must be possible with the mechanism of dispersive waves trapping by reducing the OH absorption peak, which can be achieved by a careful cleaning treatment during the fiber fabrication process, and/or by enhancing the fiber nonlinearity.

5. White-light continuous-wave supercontinuum

5.1 Benefit of GeO$_2$ doping

As explained above, CW SC generation is intimately linked to the propagation of fundamental solitons generated from MI. In order to optimize the SC bandwidth, it is thus necessary to optimize the soliton self-frequency shift effect. One of the most natural solitons to do that is to use GeO$_2$-doped fibers, because this doping is well known to enhance both Kerr and Raman nonlinearities. However, in order to be usable in CW SC generation experiments, it is important that the ZDW remains slightly lower than the pump wavelength. By adjusting the microstructured cladding properties, it is possible to find some designs with greatly enhance nonlinearity, and still controlled dispersion Barviau et al. (2011). Figure 11 illustrates this. The blue line in (a) correspond to the GVD curve and nonlinear coefficient of a pure silica PCF with a 1060 nm ZDW, and the red line corresponds to a GeO$_2$-doped PCF with a parabolic profile and a maximum refractive index difference of 20 mol.%. The microstructured cladding parameters have been adjusted so that the doped PCF has the same ZDW of 1060 nm, but in this case, the nonlinear coefficient is enhanced by a factor of about 4 at 1064 nm. Indeed, it reached about 38 W^{-1}.km^{-1} in the GeO$_2$-doped PCF, against 10 W^{-1}.km^{-1} in the pure silica one. Moreover, Fig. 11(b) shows the enhancement of the material Raman gain due to the presence of GeO$_2$ as compared to pure silica. A GeO$_2$-doped PCF corresponding to this design has thus been fabricated, and the pure silica PCF has also been used for comparison.

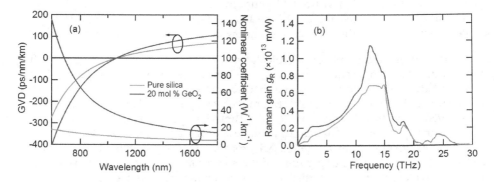

Fig. 11. (a) GVD curves (left axis) and nonlinear coefficient (right axis) calculated for a pure silica PCF (blue) and a PCF doped with a GeO$_2$ content of 20 mol.%. (b) Raman gain spectra g_R for pure silica (blue) and 20 mol.% GeO$_2$-doped silica (for a 164 nm pump).

5.2 Spectral extension to the blue

In order to highlight the benefit of using GeO$_2$-doped PCFs in the context of CW SC generation, both fibers were pumped with a CW fiber laser at 1064 nm in similar conditions. Figure 12(a) shows output spectra obtained for a pump power of 13 W and a length of 300 m for the GeO$_2$-doped PCF and 400 m for the pure silica one. The SC spectrum looks much broader in the GeO$_2$-doped PCF than in the pure silica one, yet longer.

As mentioned above, long and short-wavelength SC edges are fixed by a group-velocity matching condition between solitons and trapped dispersive waves Genty et al. (2004; 2005); Gorbach & Skryabin (2007a;b); Gorbach et al. (2006); Travers (2009); Travers & Taylor (2009). Bottom curves in Fig. 12 show group index curves of both fibers, and blue lines illustrates the group-index matching between both SC edges in the pure silica fiber. In the GeO$_2$-doped one, a dip in the spectral power density appears just below 1380 nm because of soliton accumulation just below the OH absorption band. As a consequence, a dip in the spectral power density can be observed at the corresponding group-velocity matched wavelength (around 805 nm) because of trapped dispersive wave accumulation. From the measurement of

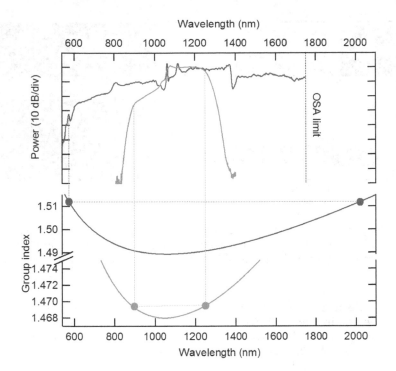

Fig. 12. Top: Supercontinuum generated in the pure silica PCF (blue line) and in the GeO$_2$-doped one (red line) for a pump power of 13 W. Bottom: corresponding calculated group index curves.

the short wavelength edge (570 nm for the GeO$_2$-doped PCF) together with the group-index matching, it is thus possible to estimate the long-wavelength one to 2040 nm (not reachable with our optical spectrum analyzer). In this case, there is thus a threefold enhancement of the SC bandwidth (in frequency) as compared to the pure silica PCF.
Note that comparable results in terms of spectral extent have been reported in pure silica PCFs Travers et al. (2008), but with a much higher pump power.

5.3 White-light generation
With the aim of still enhancing the SC bandwidth, it is possible to associate the benefits of GeO$_2$ doping and fiber tapering presented above. We have thus fabricated a GeO$_2$-doped ZDW decreasing PCF, characterized by a 50 m long uniform section followed by a 130-m long section over which the outer diameter linearly decreases from 135 to 85 μm. Figure 13(a) shows the spectrum obtained with this fiber for a 45 W pump power. It spans from 470 nm to more than 1750 nm, with an output power of 10 W. A picture of this experiment showing white-light generation is displayed in Fig. 13(b). This is the first demonstration of CW white-light supercontinuum generation.

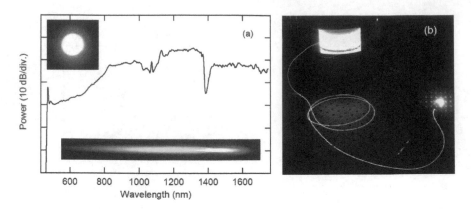

Fig. 13. (a) Supercontinuum generated in a GeO_2-doped ZDW decreasing PCF, with a pump power of 45 W. Top inset: photograph of the output beam far-field. Bottom inset: photograph of the output beam dispersed by a prism. (b) Photograph of the experiment.

6. Conclusion

It is now possible to generate continuous wave supercontinua ranging from the near ultraviolet to the near infrared with spectral power densities in the order of tens of mW/nm. Most of the manipulation of the spectra were carried out by a fine control of the fiber microstructure. These all-fiber sources are very promising for many applications requiring stable and extremely powerful sources.

7. References

Akhmediev, N. & Karlsson, M. (1995). Cherenkov radiation emitted by solitons in optical fibers, *Phys. Rev. A* 51(3): 2602–2607.

Avdokhin, A. V., Popov, S. V. & Taylor, J. R. (2003). Continuous-wave, high-power, Raman continuum generation in holey fibers, *Opt. Lett.* 28(15): 1353–1355.

Barviau, B., Vanvincq, O., Mussot, A., Quiquempois, Y., Melin, G. & Kudlinski, A. (2011). Enhanced soliton self-frequency shift and CW supercontinuum generation in GeO(2)-doped core photonic crystal fibers, *JOURNAL OF THE OPTICAL SOCIETY OF AMERICA B-OPTICAL PHYSICS* 28(5): 1152–1160.

Cumberland, B. A., Travers, J. C., Popov, S. V. & Taylor, J. R. (2008a). 29 W high power CW supercontinuum source, *Opt. Express* 16(8): 5954–5962.

Cumberland, B. A., Travers, J. C., Popov, S. V. & Taylor, J. R. (2008b). Toward visible cw-pumped supercontinua, *Opt. Lett.* 33(18): 2122–2124.

Dudley, J. M., Genty, G. & Coen, S. (2006). Supercontinuum generation in photonic crystal fiber, *Rev. Mod. Phys.* 78(4): 1135–1184.

Efimov, A., Taylor, A., Omenetto, F., Knight, J., Wadsworth, W. & Russell, P. (2003). Phase-matched third harmonic generation in microstructured fibers, *OPTICS EXPRESS* 11(20): 2567–2576.

Genty, G., Lehtonen, M. & Ludvigsen, H. (2004). Effect of cross-phase modulation on supercontinuum generated in microstructured fibers with sub-30 fs pulses, *Opt. Express* 12(19): 4614–4624.

Genty, G., Lehtonen, M. & Ludvigsen, H. (2005). Route to broadband blue-light generation in microstructured fibers, *Opt. Lett.* 30(7): 756–758.

Gonzalez-Herraez, M., Martin-Lopez, S., Corredera, P., Hernanz, M. & Horche, P. (2003). Supercontinuum generation using a continuous-wave Raman fiber laser, *Opt. Commun.* 226(1-6): 323–328.

Gorbach, A. V. & Skryabin, D. V. (2007a). Light trapping in gravity-like potentials and expansion of supercontinuum spectra in photonic-crystal fibres, *Nature Photon.* 1(11): 653–657.

Gorbach, A. V. & Skryabin, D. V. (2007b). Theory of radiation trapping by the accelerating solitons in optical fibers, *Phys. Rev. A* 76(5).

Gorbach, A. V., Skryabin, D. V., Stone, J. M. & Knight, J. C. (2006). Four-wave mixing of solitons with radiation and quasi-nondispersive wave packets at the short-wavelength edge of a supercontinuum, *Opt. Express* 14(21): 9854–9863.

Gordon, J. P. (1986). Theory of the soliton self-frequency shift, *Opt. Lett.* 11(10): 662–664.

Kobtsev, S. & Smirnov, S. (2005). Modelling of high-power supercontinuum generation in highly nonlinear, dispersion shifted fibers at CW pump, *Opt. Express* 13(18): 6912–6918.

Kudlinski, A., George, A., Knight, J., Travers, J., Rulkov, A., Popov, S. & Taylor, J. (2006). Zero-dispersion wavelength decreasing photonic crystal fibers for ultraviolet-extended supercontinuum generation, *Opt. Express* 14(12): 5715–5722.

Kudlinski, A. & Mussot, A. (2008). Visible cw-pumped supercontinuum, *Opt. Lett.* 33(20): 2407–2409.

Mussot, A., Beaugeois, M., Bouazaoui, M. & Sylvestre, T. (2007). Tailoring CW supercontinuum generation in microstructured fibers with two-zero dispersion wavelengths, *Opt. Express* 15(18): 11553–11563.

Nishizawa, N. & Goto, T. (2002). Pulse trapping by ultrashort soliton pulses in optical fibers across zero-dispersion wavelength, *Opt. Lett.* 27(3): 152–154.

Omenetto, F., Taylor, A., Moores, M., Arriaga, J., Knight, J., Wadsworth, W. & Russell, P. (2001). Simultaneous generation of spectrally distinct third harmonics in a photonic crystal fiber, *OPTICS LETTERS* 26(15): 1158–1160.

Persephonis, P., Chernikov, S. & Taylor, J. (1996). Cascaded CW fibre Raman laser source 1.6-1.9 μm, *Electron. Lett.* 32(16): 1486–1487.

Prabhu, M., Kim, N. & Ueda, K. (2000). Ultra-broadband CW supercontinuum generation centered at 1483.4 nm from Brillouin/Raman fiber laser, *Jpn. J. Appl. Phys. Part 2 - Letters* 39(4A): L291–L293.

Skryabin, D., Luan, F., Knight, J. & Russell, P. (2003). Soliton self-frequency shift cancellation in photonic crystal fibers, *Science* 301(5640): 1705–1708.

Stone, J. M. & Knight, J. C. (2008). Visibly "white" light generation in uniform photonic crystal fiber using a microchip laser, *Opt. Express* 16(4): 2670–2675.

Travers, J. C. (2009). Blue solitary waves from infrared continuous wave pumping of optical fibers, *Opt. Express* 17(3): 1502–1507.

Travers, J. C., Rulkov, A. B., Cumberland, B. A., Popov, S. V. & Taylor, J. R. (2008). Visible supercontinuum generation in photonic crystal fibers with a 400W continuous wave fiber laser, *Opt. Express* 16(19): 14435–14447.

Travers, J. C. & Taylor, J. R. (2009). Soliton trapping of dispersive waves in tapered optical fibers, *Opt. Lett.* 34(2): 115–117.

Tse, M., Horak, P., Poletti, F., Broderick, N., Price, J., Hayes, J. & Richardson, D. (2006). Supercontinuum generation at 1.0 mu m in holey fibers with dispersion flattened profiles, *Opt. Express* 14(10): 4445–4451.

Vanholsbeeck, F., Martin-Lopez, S., Gonzalez-Herraez, M. & Coen, S. (2005). The role of pump incoherence in continuous-wave supercontinuum generation, *Opt. Express* 13(17): 6615–6625.

5

Slow Light in Optical Fibers

Shanglin Hou[1] and Wei Qiu[2]
[1]School of Science, Lanzhou University of Technology, Lanzhou,
[2]Department of Physics, Liaoning University, Shenyang
China

1. Introduction

The group velocity at which light pulses propagate through a dispersive material system is very different from the vacuum speed of light c, One refers to light as being "slow" for $v_g \ll c$ (Boyd & Gauthier, 2009) or "fast" for $v_g > c$ or $v_g < 0$ (Stenner et al, 2003). For $v_g < 0$, the pulse envelope appears to travel backward in the material (Gehring et al, 2006), and hence it is sometimes referred to as "backward light."

The subject of slow light has caused keen interest in the past decade or more, and it is possible to control the group velocity of light pulses in the dispersive materials. Interest in slow and fast light dates back to the early days of the 20th century. Sommerfeld and Brillouin (Sommerfeld & Brillouin, 1960) were intrigued by the fact that theory predicts that v_g can exceed c, which leads to apparent inconsistencies with Einstein's special theory of relativity. Experimental investigations of extreme propagation velocities were performed soon after the invention of the laser (Faxvog and et al, 1970). In 1999, Harris's group research work greatly stimulated researchers' interests, which showed that light could be slowed down to 17m/s. The result was obtained in ultra cold atom clouds with the use of electromagnetically induced transparency (EIT), which induces transparency in a material while allowing it to retain strong linear and nonlinear optical properties (Hau et al, 1999). Slow light can also be obtained through the use of the optical response of hot atomic vapors (Philips et al, 2001). These early research works require hard conditions and the slow light cannot operate in room temperature.

Recently, researchers found ways to realize slow light operating in room temperature and solid-state materials, which are more suited for many practical applications, namely slow light via stimulated Brillouin scattering(SBS), slow light via coherent population oscillations (CPO), tunable time delays based on group velocity dispersion or conversion/ dispersion(C/D), slow light in fiber Bragg gratings and so on. In this chapter, we describe some of the physical mechanisms that can be used to induce slow and fast light effects in room-temperature solids (Bigelow et al, 2003) and some of the exotic propagation effects that can thereby be observed. We also survey some applications of slow and fast light within the fields of quantum electronics and photonics.

2. Fundamentals of slow and fast light

Slow and fast light refer to the group velocity of a light wave. The group velocity is the velocity most closely related to the velocity at which the peak of a light pulse moves through an optical dispersive material (Milonni, 2005), and is given by the standard result

$$v_g = \frac{c}{n_g}, \qquad n_g = n + \omega \frac{dn}{d\omega} \tag{1}$$

where n is the refractive (phase) index and ω is the angular frequency of the carrier wave of the light field. One refers to light as being slow light for $v_g \ll c$, fast light for for $v_g > c$, and backwards light for $v_g < 0$ or v_g is negative. Extreme values of the group velocity invariably rely on the dominance of the second contribution to the group index of Equation (1). This contribution of course results from the frequency dependence of the refractive index, and for this reason extreme values of the group velocity are usually associated with the resonant or near-resonant response of material systems.

According to this theory, slow light is expected in the wings of an absorption line and fast light is expected near line center, and the spatial dispersion, that is, the non-locality in space of the medium response, is another mechanism that can lead to slow light, as has been predicted (Kocharovskaya et al,2001) and observed (Strekalov et al, 2001). We catalogue the main methods to realize slow light in room temperature solid, namely slow light via stimulated Brillouin scattering(SBS), slow light via coherent population oscillations (CPO), tunable time delays based on group velocity dispersion or conversion/ dispersion(C/D), slow light in fiber Bragg gratings and so on. We will describe CPO and SBS slow light in more details.

3. Slow light via coherent population oscillations (CPO)

3.1 Introduction of CPO

The technique of coherent population oscillation(CPO) is also exploited to reduce the group velocity.The process of CPO allows the reduction of absorption and simultaneously provides a steep spectral variation of the refractive index which leads to a strong reduction of the optical group velocity,i.e.,slow light propagation.This process is easily achieved in a two-level system which interacts with a signal whose amplitude is periodically modulated.The population of the ground state of the medium will be induced to oscillate at the modulation frequency. This oscillation creates an arrow hole in the absorption spectrum, whose linewidth is proportional to the inverse of the relaxation lifetime of the excited level. CPO is highly insensitive to dephasing processes in contrast to what happens in other schema such as EIT, where the width of the spectral hole burned in the absorption profile is proportional to the inverse of the dephasing time of the ground state. That makes CPO an appropriate technique to easily achieve slow light propagation in solid-state materials at room temperature.

3.2 Theoretical mode

Making use of this technology, we observe optical pulse delay and advancement propagation in an erbium-doped optical fiber. Compared to other solid material, erbium-doped optical fiber allows for long interaction lengths, which can be desirable in producing strong influence(Schwartz& Tan, 1967). We obtain the controllable pulse delay continuously from positive to negative by using a separate pump laser.

When pumped at 980nm, the erbium-doped fiber acts as a three-level molecular system. The relaxation time from the metastable state to the ground state is much greater than the one from the excited state to the metastable state. The energy levels and pumping scheme that we employed to observe slow and fast light in erbium-doped fiber is shown in Fig 1.

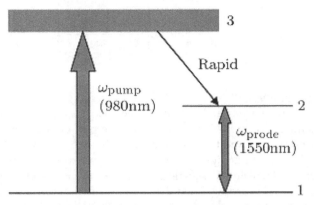

Fig. 1. The energy levels and pumping scheme we employed to observe slow and fast light in erbium-doped fiber.

The relaxation time from the excited state associated with the signal frequency to the ground state is much greater than the one from the excited state associated with the pump frequency to the excited state associated with the signal frequency. The population of upper excited level is approximately equal to zero, which indicates $n_1 + n_2 = \rho$. The population density of the ground state will accord with the rate equation(Novak & Gieske, 2002)

$$\frac{\partial n_1}{\partial t} = -R_{13}n_1 - W_{12}n_1 + W_{21}(\rho - n_1) + \frac{\rho - n_1}{T_1}, \tag{2}$$

where W denotes transition rates associated with signal and R presents transition rates associated with pump. T_1 is the lifetime of the excited state and ρ is the erbium ion density. The transition rates are also functions of t and z, as well as being proportional to the pump and signal powers. According to the equations for the transition rates and neglecting the losses through the fiber, we have

$$\frac{\partial}{\partial t}N_1 = I_p(L,t) - I_p(0,t) + I_s(L,t) - I_s(0,t) + \frac{N_0 - N_1}{T_1}. \tag{3}$$

If we modulate the signal(~1550nm) intensity as

$$I(0,t) = I_s^0(0)(1 + \mu_s \cos \delta t). \tag{4}$$

here, $I_s^0(0)$ is the average input power at the input ($Z=0$) and $I_s^0(0)\mu_s = I_m(0)$ is the modulation amplitude. A single intensity-modulated beam contains only a carrier wave (to act as the pump) and two sidebands (to act as probes) on the output spectrum, which induce population oscillation. The population of the ground state is given by

$$N_1(t) = N_1^0\left[1 + \xi \cdot \cos(\delta t + \phi)\right], \tag{5}$$

where N_1^0 is the mean (un-modulated) steady-state population. We next find the steady-state solution to Eq. (3). Finally, We can determine the expression of the delay of the optical signal

$$\Delta t = \frac{1}{\delta \times 2\pi} \arctan\left(\frac{\delta}{\frac{\delta^2_{\text{eff}} + \delta^2}{\eta} - \delta_{\text{eff}}}\right).$$ (6)

3.3 Experimental results

The signal optical field from a distributed feedback laser diode operating at 1550nm through the attenuator is divided into two parts: one part of laser (98%) goes through an erbium-doped optical fiber and then to an InGaAs photodetector with 10MHz bandwidth. The other part of laser output signal (2%) is sent directly to an identical photodetector to be used as reference.Transmitted signals are received by photodetectors, and sent into a digital oscillograph for recording. Then the comparison between the reference signal and the EDOF signal is made in a computer (Sargent,1978, Boyd & Gauthier, 2005). The group velocity in fibers can be inferred. In the experiments, the injection current of the laser is sinusoidally modulated by a function generator. We use single mode, Al_2SiO_5-glass-based erbium-doped optical fibers at several ions density. The experimental setup is shown in Fig.2.

The absorption coefficient α and the emission coefficient β are related to their cross sections respectively and shown by the following

$$\alpha_s = \Gamma_s \sigma_{12} \rho \quad \alpha_p = \Gamma_p \sigma_{13} \rho \quad \beta_s = \Gamma_s \sigma_{21} \rho .$$ (7)

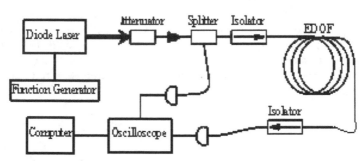

Fig. 2. The experimental setup used to observe slow light in an erbium-doped optical fiber.

Parameters used for the calculation are α_s =31.71 dB/m, α_p =42.3dB/m, β_s =47.665 dB/m, T_1 = 10.5 ms, L=2m, and ρ =6.3×10^{25}. Our experimental results are obtained through use of modulation techniques such that the optical field contains only a carrier wave (to act as the pump)and two sidebands (to act as probes). Because the decay time is so long (about 10.5ms), this oscillation will only occur if the beat frequency (δ) between the pump and probe beams is small so that δT_1 ~1.When this condition is fulfilled, the pump wave can efficiently scatter off the temporally modulated ground state population into the probe wave, resulting in reduced absorption of the probe wave. We consider the Kramers-Kronig relations, which show that a narrow hole in an absorption spectrum will produce strong normal dispersion. Fig.3 illustrates the results of our experiments.

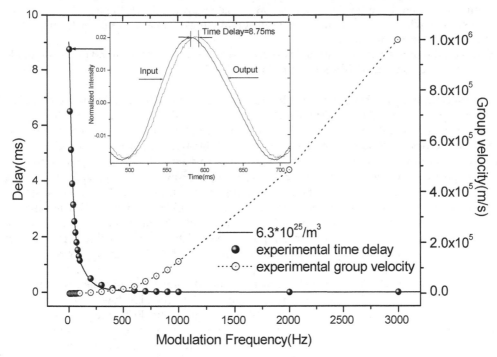

Fig. 3. Observed time delay as a function of the modulation frequency for input power of 1.85mW. The solid line is the theoretical fit to the experimental data. The open circles represent the measured group velocity. The inset shows the normalized 5Hz input (solid line) and output (dashed line) signal. The signal is delayed 8.75ms corresponding to a group velocity as low as 228.57m/s.

In Fig.3, we show the measured delay in an optical fiber with erbium ion density of $6.3 \times 10^{25}m^{-3}$ and compare it with the numerical solution of Eq.(6) for input power of 1.85mW. We observe the largest delay, 8.75ms, which corresponds in the inset of Fig.3. The inferred group velocity is as low as 228.57m/s. A maximum fractional delay of 0.129 is observed at the modulation frequency of 60Hz. Fig.3 shows the delay as a function of modulation frequency in the low frequency region.

4. Slow light via stimulated Brillouin scattering (SBS)

4.1 Introduction
Slow light based on stimulated Brillouin scattering (SBS) in optical fibers has attracted much more interests for its potential application in optical buffering, data synchronization, optical memories and optical signal processing. Compared with previously demonstrated slow-light techniques (Gehring et al, 2008, Zhu et al, 2007), such as electromagnetically induced transparency (EIT) (Hau et al, 1999) and coherent population oscillations (CPO) (Bigelow et al, 2003), it has a lot of advantages, for instance, the simple, flexible and easy-to-handle SBS can be realized in room temperature; the optical fiber components based on it can easily integrated with the existing telecommunications infrastructure; the slow-light resonance can

be tunable within the optical communications wavelength windows; the use of optical fiber allows for a relaxed pump-power requirement owning to long interaction length, small effective mode area and so on.

However, the SBS-induced group index change is always so small in standard single mode fiber and dispersion shift fibers (DSFs) (Song et al, 2005) to delay the time very little. In order to explore suitable optical fibers served as slow light generation with much efficiency, some special optical fibers, such as chalcogenide fiber (Abedin,2005, Song et al, 2006), tellurite fiber (Abedin,2006), bismuth fiber (Jauregui et al, 2006) and so on, have been extensively studied, these kinds of optical fiber are usually with large gain coefficient and low loss coefficient. Though long pulse delay can be obtained using cascaded fiber segments joined by unidirectional optical attenuators to overcome pump depletion (gain saturation) and amplified spontaneous Brillouin emission (ASBE), it's always accompanied with serious pulse distortion (Song et al, 2005). So gain tailoring is used in pulse distortion management to keep a balance between time delay and pulse distortion (Stenner et al, 2007). To overcome the narrow band spectral resonance of SBS which limits the maximum data rate of the optical system, a simple and inexpensive pump spectral broadening technique is used in broadening the SBS slow light bandwidth (Herraez et al, 2006), which paves the way towards real applications based on SBS slow light.

Numerical studies of SBS slow light focusing on different pulse parameters were also studied (Kalosha et al, 2006), which provide an insight into the SBS slow light process, but we can't learn a lot about how the optical fiber structures and Brillouin gain parameters influence on the SBS process, the time delay and the pulse shape. In this section, the SBS model in optical fiber is described and the three coupled SBS equations are solved by the method of finite difference with prediction-correction, the effects of gain coefficient, gain bandwidth and effective mode area on time delay and pulse broadening are demonstrated. Considering the injected stokes pulse shape, the influence of its sharpness, magnitude and duration on delay time and pulse broadening factor was observed mainly, and its reason was analyzed. These results provide base for designing optical buffer, time delay line or other optical components based on the SBS slow light technologies.

4.2 Theory foundation and numerical model

The process of SBS is the interaction of two counter-propagating waves, a strong pump wave and a weak Stokes wave. If a particular frequency relation is satisfied

$$v_{pump} = v_{Stokes} + v_B , \tag{8}$$

Where v_{pump} and v_{Stokes} are the frequency of pump wave and stocks wave respectively, v_B is the Brillouin frequency. Then an acoustic wave is generated which scatters photons from the pump to the Stokes wave and the interference of these two optical waves in turn stimulates the process. From a practical point of view, the process of SBS can be viewed as a narrowband amplification process, in which a continuous-wave pump produces a narrowband gain in a spectral region around $v_{pump} - v_B$. In this paper, the Stokes pulse is set on the SBS gain line center to achieve the maximum delay.

For simply describing the SBS process, assume: (1) Transverse field variations are neglected, Stokes and pump fields are assumed to vary with time t and space z only. (2) The slowly varying envelope approximation (SVEA) SBS model is used, i.e., the field amplitudes are assumed to vary slowly in time and space as compared with their temporal and spatial frequencies. (3) The initial (t=0) phonon field is zero and the Stokes output grows from an injected Stokes field at z=0. (4) The frequency difference between the pump and Stokes wave is set to the Brillouin shift of the fiber, i.e., the Stokes pulse is on the SBS line center.

Considering a Brillouin amplifier where the pump wave counter-propagates through the fiber with respect to the Stokes pulse, the SBS process can be described by one-dimensional coupled wave equations involving a backward pump wave ($-z$ direction), a forward Stokes wave ($+z$ direction), and a backward acoustic wave. Under the slowly varying envelope approximation (SVEA) and neglecting the transverse field variations, the equations are written as follows (Damzen et al,2003)

$$-\frac{\partial A_p}{\partial z}+\frac{n}{c}\frac{\partial A_p}{\partial t}=-\frac{\alpha}{2}A_p+ig_2A_sQ, \tag{9}$$

$$\frac{\partial A_s}{\partial z}+\frac{n}{c}\frac{\partial A_s}{\partial t}=-\frac{\alpha}{2}A_s+ig_2A_pQ^*, \tag{10}$$

$$\frac{\partial Q}{\partial t}+\frac{\Gamma_B}{2}Q=ig_1A_pA_s^*, \tag{11}$$

where A_p, A_s, and Q are the amplitudes of the pump wave, the Stokes wave, and the acoustic wave, respectively; n is the group refractive index when SBS is absent; α is the loss coefficient of the fiber; $\Gamma_B/2\pi$ is the bandwidth (FWHM) of the Brillouin gain; g_1 is the coupled coefficient between the pump wave and the Stokes wave, g_2 is the coupled coefficient between the pump (Stokes) wave and the acoustic wave, $g_0=4g_1g_2/\Gamma_B$ is the peak value of the Brillouin gain coefficient.

According to the small signal steady state theory of stimulated Brillouin scattering, the pump power $P_{critical}$ required to reach Brillouin threshold in a single pass scheme is related to the Brillouin gain coefficient g_0 by the following equation:

$$g_0(P_{critical}/A_{eff})L_{eff}\cong21, \tag{12}$$

where $P_{critical}$ is the power corresponding to the Brillouin threshold, L_{eff} is the effective length defined as $L_{eff}=\alpha^{-1}[1-\exp(-\alpha L)]$, from Eq.(12) we can obtain the threshold pump intensity

$$I_{critical}=P_{critical}/A_{eff}\cong21/(g_0L_{eff}). \tag{13}$$

Once reaching the threshold pump intensity, a large part of the pump power is transferred to the Stokes wave, resulting in the generation of Stokes wave at the output depletes the pump seriously and leads to serious Stokes pulse distortion. In our simulations, we consider the pump intensity is near the Brillouin threshold and obtain the Stokes gain around 16 using the previous parameters, here the Stokes gain is defined as:

$$Gain=\log\left(\frac{P_{out}}{P_{in}}\right), \tag{14}$$

where P_{out} and P_{in} are the output and input of the Stokes power, respectively.
Let us assume that pump wave is continuous wave and stokes field is sufficiently weak. The group index is the function of frequency described as follow(Zhu et al, 2005)

$$n(\omega_s) = n_g + \frac{cg_B I_p}{\Gamma_B} \frac{1 - 4\delta\omega^2 / \Gamma_B^2}{(1 + 4\delta\omega^2 / \Gamma_B^2)^2} \tag{15}$$

where I_p is the optical intensity of pump wave; $\delta\omega$ is the margin between the angular frequency of stokes pulse and the center angular frequency of the gain bandwidth; $g_B = \dfrac{\gamma_e^2 \omega_p^2}{n_g v_B c^3 \rho_0 \Gamma_B}$ is the line-center gain factor which is associated with the material physical properties, v_B is the velocity of acoustic wave.

Delay time T_d is defined to describe the difference of the arrival time when the output stokes pulse reach its maximum between when SBS occurs and doesn't occurs, $T_{rd} = T_d / T$ to describe the relative delay time with T which is the FWHM of the injected stokes pulse. According to the weak signal theory, delay time and B are given by (Velchev et al, 1999)

$$T_d = G / \Gamma_B \tag{16}$$

$$B = \sqrt{1 + \frac{16\ln 2}{T^2 \Gamma_B^2} G} \tag{17}$$

where $G = g_B I_p L$ is the weak signal gain parameter; L is the fiber length. For the purpose of indicating how much the pump wave energy contributes to the stokes wave energy, we define real gain as

$$G_r = \log(P_{out} / P_{in}) \tag{18}$$

where P_{out} and P_{in} are power of the output and input stokes wave, respectively.
We assume that injected stokes pulse is super-Gaussion shaped

$$U(t) = \exp\left[-\frac{1}{2}\left(\frac{t}{T_0}\right)^{2m} \right] \tag{19}$$

where t is the time of pulse transmission, U(t) is normalized amplitude, T_0 is the half width of pulse (at $1/e$-intensity point). The parameter m controls the degree of edge sharpness (for m=1, it is Gaussion-shaped; for m>1, it is super-gaussion-shaped, and the degree of edge sharpness is increased with m). For different m, pulse with same FWHM can be written as

$$U(t) = \exp\left[-2^{2m}(\ln 2)\left(\frac{t}{T}\right)^{2m} \right] \tag{20}$$

Furthermore, we define average intensity of normalized super-Gaussion pulse with FWHM of T as

$$\bar{I}_s = \frac{1}{T} \int_{-T/2}^{T/2} (U(t))^2 \, dt = \frac{1}{T} \int_{-T/2}^{T/2} \left(\exp\left(-2^{2m} (\ln 2) \frac{t^{2m}}{T^{2m}} \right) \right)^2 dt \qquad (21)$$

In our numerical processing, applying the slowly-varying envelope approximation (SVEA) for both pump and stokes fields, firstly, we obtained the values of ρ at some time by solving the Eq.(11) with Fourier transformation and inverse Fourier transformation. Secondly, we transform the Eqs (9-10) into single variable partial differential equations by using characteristics. Finally, applying the value of ρ we obtained to the single variable partial differential equations, we can calculate the values of E_p and E_s at next time by using the fourth-order Runge-Kutta formula. And we set these results as the initialization value to achieve the value of ρ at next time. Repeating above steps, we can achieve the output stokes pulse at anytime.

4.3 Numerical simulation results
In order to study the situation where the pump is depleted, we solve the Eqs. (9)-(11) numerically using the method of implicit finite difference with prediction-correction to determine how gain coefficient, gain bandwidth and effective mode area influence SBS slow light.

In our simulation, the parameters are considered from the common single-mode fiber, and select: fiber length L=25m, pump wavelength λ =1550nm, group refractive index n=1.45, effect mode area A_{eff} =50 μm^2, loss coefficient α =0.2dB/km, gain bandwidth (FWHM) $\Gamma_B / 2\pi$ =40MHz, gain coefficient $g_0 = 5 \times 10^{-11}$ m/W. We assume the pump wave is CW and the Stokes wave is Gaussian shaped with the peak power of 0.1 μW and the FWHM pulse width of 120ns (its FWHM bandwidth in frequency domain is around 3.7MHz which is much smaller than that of SBS gain bandwidth we use).

4.3.1 Influence of gain coefficient on time delay and pulse broadening
The curves of the pulse delay and pulse broadening factor as a function of the gain with different gain coefficient g_0 were shown in Fig.4. It can be seen from Fig.4(a) that the time delay increases linearly with Stokes gain when the gain is small ($\leq$10), that's because the pump isn't completely affected when the gain is small. For larger gain, pump depletion becomes more and more seriously, the time delay increases slowly with gain and reaches its maximum before decreasing with gain. At the same time, for larger gain coefficient, the time delay decreases with increasing gain more quickly and even leads to pulse advancement which can be explained by gain saturation. It can also be seen that the smaller gain coefficient reaches the gain saturation at a larger gain and the maximum time delay is accordingly larger, the gain saturation limits the maximum time delay for a Stokes pulse at a given input power.

The pulse broadening factor for different gain coefficients as a function of gain was shown in Fig.4(b). It shows that the pulse broadening factor is also increasing linearly with gain when the small signal regime holds. As the gain further increases, the pulse broadening factor increases slowly with gain and then gradually decreases to less than 1, it means that the pulse become more and more narrower, the pulse with larger gain coefficient narrows more seriously than the smaller one. The time delay is always accompanied with pulse distortion, the Stokes pulse broadens a little in the small signal regime but can narrow

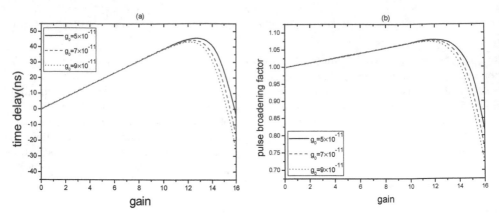

Fig. 4. (a) Time delay and (b) pulse broadening as a function of gain with different gain coefficients.

largely in the gain saturation regime. Fig.5 shows the normalized output pulse shapes with the gain coefficient $g_0 = 5 \times 10^{-11}$ m/W at gain=0, 12, and 17, respectively. The output Stokes pulse with a maximum time delay ~45 ns at gain=12 with a little distortion while the output Stokes pulse is advanced by 42.9 ns at gain=17 but is distorted substantially.

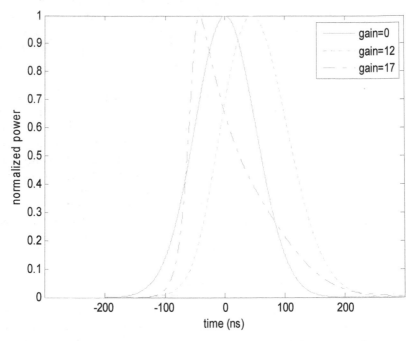

Fig. 5. Normalized output Stokes pulse at gain=0, 12, 17 with the gain coefficient $g_0 = 5 \times 10^{-11}$ m/W.

Next, we consider the time delay and pulse broadening factor varying with the gain coefficient at a given pump peak power 0.125W shown in Fig.6. From Fig.6(a) we can see that the time delay increases with the increasing gain coefficient in a linear fashion. Fig.6(b) shows that the pulse broadening factor also increases with the increasing gain coefficient. Note that the maximum gain parameter is 6.25 at the gain coefficient $g_0 = 1 \times 10^{-10}$ m/W, which satisfies the small signal condition.

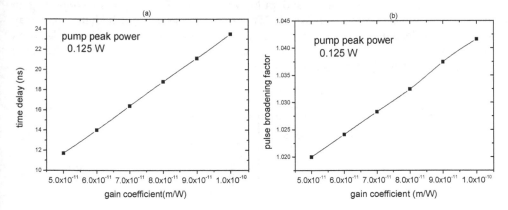

Fig. 6. (a) Time delay and (b) pulse broadening as a function of gain coefficient at a given pump power.

4.3.2 Influence of gain bandwidth on time delay and pulse broadening

Fig.7(a) shows the time delay as a function of gain with different gain bandwidths. The time delay increases with the gain linearly when the gain is small, but for smaller gain bandwidth, the time delay increases with the gain more quickly and reaches the saturation at a larger gain, the maximum time delay is accordingly larger. Once reaching the gain saturation, the time delay also decreases more quickly for smaller gain bandwidth.

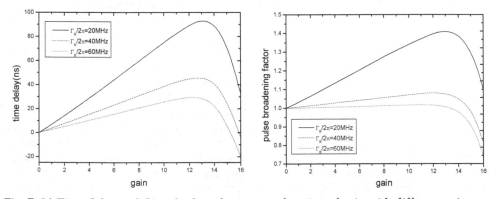

Fig. 7. (a) Time delay and (b) pulse broadening as a function of gain with different gain bandwidths.

The pulse broadening factor as a function of gain with different gain bandwidths was shown in Fig.7(b). It can be seen that the pulse broadening factor increases with the gain before gain saturation and then it decreases with the increasing gain which is similar with time delay versus gain in Fig.7(a). The smaller the gain bandwidth is, the more quickly the broadening factor increases with the gain in the small signal regime and decreases with the gain in the gain saturation. It indicates that the pulse with smaller gain bandwidth always obtains the longer time delay but at the cost of much larger pulse broadening.

We also calculated the time delay and pulse broadening factor as a function of gain bandwidth at a given pump peak power 0.125W. As we can see in Fig.8, both the time delay and pulse broadening factor decrease with the increasing gain bandwidth and keep an inverse proportion to it. The maximum gain parameter is 6.25, which is also in the small signal regime for these different gain bandwidths.

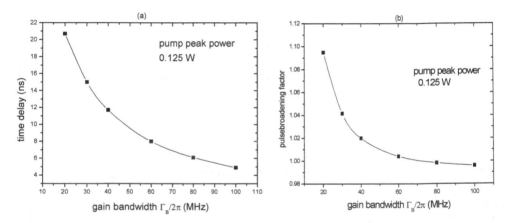

Fig. 8. (a) Time delay and (b) pulse broadening as a function of gain bandwidth at a given pump power.

4.3.3 Influence of effective mode area on time delay and pulse broadening

For the pulse with same peak power, it has a larger intensity for the smaller effective mode area, which increases its intensity in the other way, so it can also influence the gain saturation obviously. As can be seen from Fig.9(a), in the small signal regime, the time delay still increases with the gain linearly for different effective mode areas, the pulse with smaller effective mode area reaches the gain saturation at a smaller gain, and the maximum time delay is accordingly smaller. Once reaching the gain saturation, the pulse with smaller effective mode area also decrease more quickly than the others.

Fig.9(b) shows the pulse broadening factor versus gain, the pulse broadening factor increases linearly with the increasing gain in the small signal regime, which is the same as the time delay versus gain. As has been said before, the pulse with larger effective mode area reaches the gain saturation at a larger gain and its maximum pulse broadening factor is accordingly larger. In the gain saturation regime, the pulse with smaller effective mode area narrows more seriously than the others at a fixed gain. We can even see that the pulse broadening factor begins to increase for the gain around 16.

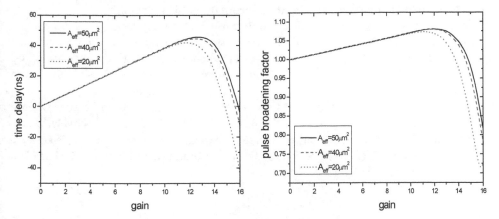

Fig. 9. (a) Time delay and (b) pulse broadening as a function of gain with different effective mode areas.

We also investigate the time delay and pulse broadening factor as a function of effective mode area at a given pump power 0.125W. It can be seen from Fig.10, both the time delay and pulse broadening factor decrease with the increasing effective mode area and keep an inverse proportion to it. As previously mentioned, we also make sure that the corresponding gain parameter is within the small signal regime for these different effective mode areas.

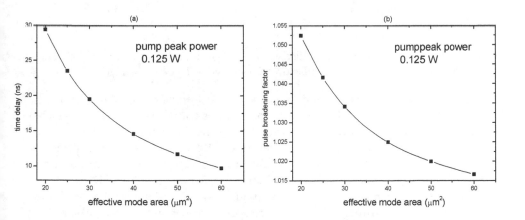

Fig. 10. (a) Time delay and (b) pulse broadening as a function of effective mode area at a given pump power.

4.3.4 The influence of stokes pulse with different m on SBS slow light

We first consider the pulse time delay and pulse broadening factor as a function of parameter real gain for super-Gaussion-shaped pulse with different m, which is indicated in Fig.11. In this case, T=120ns, $P_{in} = 0.1 \mu W$. Fig.11 (a) shows that with the increase of G_r, T_d increases accordingly and reaches its maximum. Then it decreases with further increasing G_r, even becomes negative. Comparing with different m, we can see that maximum G_r and the time when maximum G_r obtains decrease with m because when m changes from 0.5 to 3, $\bar{I}s$ is equal to 0.5410, 0.6805, 0.7559 and 0.8705, respectively, i.e., for the same duration and apex, the super-Gaussion-shaped pulse with higher m is easier to reach gain saturation. And T_d when T_d reaches its maximum will decrease, which leads to the reducing of maximum T_d.

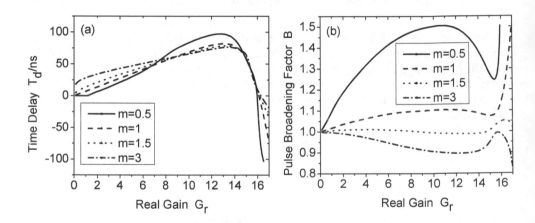

Fig. 11. Parameters of output stokes pulse versus real gain for different parameter m. a) Time delay T_d versus real gain; b) Pulse broadening factor B versus real gain.

We observe an advantageous phenomenon for practical applications. When m=1.5, pulse broadening factor B is close to 1. While m=3, B decreases with increasing G_r and reaches its peak value at gain=13. Then it increase with G_r and reaches its maximum at G_r =15.5, it decrease with further increasing G_r. The reason why B has the rule can be explained well by Fig.12. Considering SBS process and no SBS process, respectively, for three different m, the normalized output stokes pulses at G_r =5 and G_r =13 are shown in Fig.12, where t is the time axis. It is indicated from Fig.12 (a) that with increasing m the leading edge and the trailing edge of super-Gaussion-shaped pulse become steeper and steeper in time domain. So they will become broader and broader in the frequency domain. Considering equation (2), we can conclude that the difference among the speeds of points at the leading edge will increase and the leading edge will be compressed more. Moreover, the increasing m leads to the increasing energy of pulse with same peak value. And most energy of pump wave is depleted in the leading edge of stokes pulse, the trailing edge only can get less energy from pump wave, the broadening of the trailing edge of stokes pulse is limited. All of this can contribute to decreasing B and result in B is almost close to 1 before stokes pulse is near saturation. When G_r increases to saturation gain step by step, the trailing edge gets more and more energy, resulting in broadening of the trailing, i.e., B will increase, like shown in Fig.12(b). When G_r go on increasing, it is out of the range of weak signal, the output stokes pulse is distortion seriously.

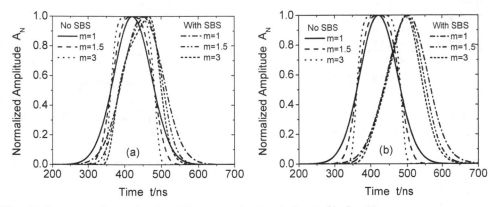

Fig. 12. Output stokes pulse for different real gain. a) G_r =5; b) G_r =13.

4.3.5 The influence of stokes pulse with different power and FWHM on SBS slow light

Based on the result of above that when m=1.5 B is very close to 1, the next numerical simulation will select different power and duration of super-Gaussion pulse with m=1.5 as the injected stokes pulse. Fig.13 shows that delay and B as a function of parameter G_r for super-Gaussion-shaped pulse with different power. It can be seen form Fig.13 (a) that maximum T_d and G_r needed for obtaining maximum T_d increase with injected power. And before entering gain-saturation regime T_d is equal to each other. The reason is in the condition of weak signal delay is in direct proportion to gain approximately. As we can see from Fig.13(b) that when G_r is smaller than saturation gain, B is close to 1 and B of the pulse which has the largest power will reach the peak shown in Fig.11(b) firstly with increasing G_r. The peak value becomes lager and larger, which correspond to high power pulse is easy to enter saturation regime.

Fig.14 shows that T_{rd} and B as a function of parameter G_r for super-Gaussion-shaped pulse with different T. The power of injected stokes pulse is 0.1 μW and other parameter is the same like above. It is indicated that smaller duration pulse can obtain lager relative delay. The G_r needed for obtaining maximum T_{rd} becomes lager and larger with the decreasing T. The reason is for same peak value the smaller duration pulse contains less energy. Then when it enters gain-saturation regime it need more energy and higher saturation gain results in higher T_{rd}. The changing rule of pulse broadening factor of the pulse which duration is less than 120ns is different from the one which duration is 120ns, however, it is the same with the one when injected stokes pulse is Gaussion-shaped. The main reason is the energy

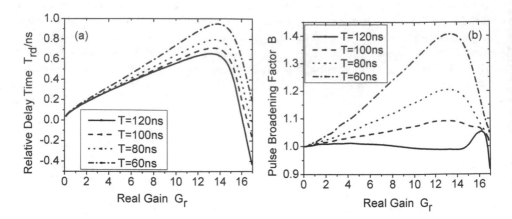

Fig. 13. Parameters of output stokes pulse versus real gain for different power. a)Time delay T_d versus real gain; b) Pulse broadening factor B versus real gain.

which pulse contains decrease with the decreasing duration for same peak value. The energy getting from pump wave decreases, too. So the tailing edge of stokes pulse can obtain more energy than the one when T=120ns, resulting in the tailing edge broaden widely. This counteracts the compression corresponding to the steep leading edge. It can be predicted that increasing m can make the leading edge steeper and can decrease B. Our numerical result proved it. For the pulse with T=60ns, B at maximum T_{rd} decreases from 1.406 to 1.295 when m is changed from 1.5 to 5.

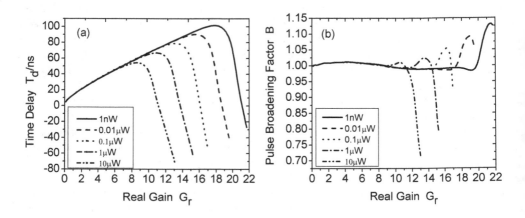

Fig. 14. Parameters of output stokes pulse versus real gain for different duration. a) Relative time delay T_{rd} versus real gain; b) Pulse broadening factor B versus real gain.

5. Conclusion

We make a numerical study of the SBS slow light in optical fibers, and consider the influences of gain coefficient, gain bandwidth and effective mode area on time delay and pulse broadening. In the small signal regime, we find that the time delay and the pulse broadening factor increase with the increasing gain, but for pulse with the smaller gain bandwidth has a larger slope than the others. In the gain saturation regime, the pulse with larger gain coefficient, smaller gain bandwidth, smaller effective mode area begins to decrease more quickly in the gain range of 0~16. For the gain larger than 16, the pulse advancement becomes more obviously and the distortion also becomes more seriously, which may render the delay useless. We also investigate the time delay and pulse broadening factor vary with the increasing gain coefficient, gain bandwidth and effective mode area at a given pump power whose gain parameter is in the small signal regime, and find that the time delay and pulse broadening factor are proportional to the gain coefficient, whereas inversely proportional to the gain bandwidth and the effective mode area.

According to the above numerical calculation and theory analysis, we find that decreasing the power and duration of injected stokes pulse induces increasing delay time and pulse broadening factor; using super-Gaussion-shaped pulse as the injected stokes pulse can contribute evidently to decreasing pulse broadening factor in low frequency. Selecting pretty m can get perfect delay time and pulse broadening factor. Though this adjusting effect will become weaker for a shorter pulse, this reform still takes advantage of decreasing the error rate of all-optical-buffer.

6. Acknowledgment

This work was supported by the National Natural Science Foundation of China (Grant No. 61167005), the Provincial Natural Science Foundation of Gansu Province (Grant No. 1010RJZA036), the Provincial Natural Science Foundation of Guangdong Province (Grant No. 110451170003004948) and Dongguan Science and Technology Program (Grant No.2008108101002).

7. References

Boyd R.W. and Gauthier D. J.(2009), Controlling the Velocity of Light Pulses, *Science*, Vol.326,No.5956,pp.1074-1077, ISSN 0036-8075.

Stenner M. D, Gauthier D. J, Neifeld M. A(2003). The speed of information in a 'fast-light' optical medium, *Nature*, Vol. 425, pp. 695-698.ISSN0028-0836.

Gehring G. M, Schweinsberg A., BarsC. i, Kostinski N., Boyd R. W.(2006) Observation of Backward Pulse Propagation Through a Medium with a Negative Group Velocity, *Science*, Vol.312,No.5775,pp.895-897, ISSN 0036-8075.

Brillouin L(1960), *Wave Propagation and Group Velocity* ,Academic Press, New York.

Faxvog F. R, Chow C. N. Y, Bieber T., and Carruthers J. A(1970). Measured Pulse Velocity Greater Than *c* in a Neon Absorption Cell, *Appl. Phys. Lett.*, Vol.17, No.5, pp.192-193,ISSN0003-6951.

Hau L. V., Harris S. E., Dutton Z. et al(1999). Light speed reduction to 17 metres per second in an ultracold atomic gas. *Nature*, Vol.397, pp.594–598, ISSN0028-0836.

Philips D. F., Fleischhauer A., Mair A. et al(2001). Storage of light in atomic vapor. *Phys. Rev. Lett.*, Vol. 86,No.5,pp.783–786. ISSN1079-7114.

Bigelow M. S., Lepeshkin N. N., Boyd R. W(2003). Superluminal and slow light propagation in a room-temperature solid. *Science*, Vol.301,No.5630,pp.200–202, ISSN 0036-8075.

Milonni P. W(2005), *Fast Light, Slow Light, and Left-Handed Light*, Institute of Physics Publishing, Bristol, UK.

Kocharovskaya, O.; Rostovtsev, Y.; Scully, M.O(2001). Stopping light via hot atoms, *Phys. Rev. Lett.* Vol. 86,No.4,pp. 628-631, ISSN1079-7114.

Strekalov, D.; Matsko, A.B.; Yu, N.; Maleki, L.J. J(2004). Influence of inhomogeneous broadening on group velocity in coherently pumped atomic vapour, *Mod. Opt.* Vol.51, No.16,pp.2571-2578, ISSN0950-0340.

Schwartz S.E, Tan T.Y(1967). Wave Interactions in Saturable Absorbers. *Appl. Phys. Lett.* Vol. 10,No.1,pp.4~7, ISSN0003-6951.

Hillman L.W., Boyd R.W., Krasinski J., and Stroud C.R(1983). Observation of a Spectral Hole Due to Population Oscillations in a Homogeneously Broadened Optical Absorption Line. *Opt. Commun.* Vol. 45,No.6,pp. 416~419, ISSN0030-4018.

Bigelow M. S., Lepeshkin N. N., and Boyd R. W(2003). Observation of Ultraslow Light Propagation in a Ruby Crystal at Room Temperature. *Phys. Rev. Lett.* Vol.90, No.11,pp. 113903, ISSN1079-7114.

Boyd R.W(1992). *Nonlinear Optics*, Academic Press, San Diego.

A. Schweinsberg, N. N. Lepeshkin, M. S. Bigelow, R.W.Boyd, and S.Jarabo. Observation of Superluminal and Slow light Propagation in Erbium-Doped Optical Fiber. Europhys. Lett. 2006,73(2):218-224

Shin H., Schweinsberg A., Gehring G., Schwertz K., Chang H. J., and Boyd R. W(2007). Reducing Pulse Distortion in Fast-Light Pulse Propagation through an Erbium-Doped Fiber Amplifier. *Opt. Lett.*, Vol.32,No.8,pp.906-908,ISSN0146-9592.

Song K. Y., Herráez M. G., and Thévenaz L.(2005). Long Optically Controlled Delays in Optical Fibers. *Opt. Lett.*, Vol.30,No.14,pp.1782-1784, ISSN0146-9592.

Novak S., Gieske R(2002). Analytic Model for Gain Modulation in EDFAs. *J Lightwave Technol.*, Vol.20,No.6,pp.975-985,ISSN0733-8724.

Sargent III M(1978). Spectroscopic techniques based on Lamb's laser theory. *Phys. Rep.*,Vol. 43,No.5,pp. 223-265,ISSN0370-1573.

Boyd,R. W, Gauthier D. J, Gacta A.L, and Willner A.E(2005). Maximum Time Delay Achievable on Propagation through a Slow-light Medium. *Phys. Rev. A.*, Vol. 71,No.2,pp.023801-1-023801-4, ISSN1094-1622.

Gehring G. M., Boyd R. W.,. Gaeta A.L., Gauthier D. J. and Willner A. E(2008). Fiber-Based Slow-Light Technologies, *Journal of Lightwave Technology*, Nol.26,No.23,pp. 3752-3762, ISSN0733-8724.

Zhu Z., Gauthier D. J., Boyd R.W(2007). Stored Light in an Optical Fiber via Stimulated Brillouin Scattering, *Science*, Vol.318,No.5857,pp.1748-1750, ISSN 0036-8075.

Song K. Y., Herraez M. G., and Thevenaz L(2005). Observation of pulse delaying and advancement in optical fibers using stimulated Brillouin scattering, *Opt. Express*, Vol.13,No.1,pp.82-88 , eISSN1094-4087.

Abedin K. S.(2005), Observation of strong stimulated Brillouin scattering in single-mode As2Se3 chalcogenide fiber, *Opt. Express*, Vol.13,No.25,pp.10266-10271,eISSN1094-4087.

Song K. Y, Abedin K. S, Hotate K., Herraez M. G and Thevenaz L(2006), Highly efficient Brillouin slow and fast light using As2Se3 chalcogenide fiber, *Opt. Express*, Vol.14,No.13, pp.5860–5865, eISSN1094-4087.

Abedin K. S(2006), Stimulated Brillouin scattering in single-mode tellurite glass fiber, *Opt. Express*, Vol.14,No.24, pp.11766–11772, eISSN1094-4087.

Jauregui C., Ono H., Petropoulos P. and Richardson D. J(2006), Four-fold reduction in the speed of light at practical power levels using Brillouin scattering in a 2-m bismuth-oxide fiber, in *Proc. of Conference on Optical Fiber Communication (OFC 2006)*, Paper PDP2 .

Stenner M. D., Neifeld M. A., Zhu Z., Dawes A. M. C. and Gauthier D. J(2005). Distortion management in slow-light pulse delay, *Opt. Express*, Vol.13, No.25, pp. 9995-10002 , eISSN1094-4087.

Schneider T., Henker R., Lauterbach K. U., and Junker M(2007). Comparison of delay enhancement mechanisms for SBS-based slow light systems, *Opt. Express*, Vol.15,No.15),pp.9606-9613, eISSN1094-4087.

Schneider T., Henker R., Lauterbach K. U., and Junker M(2008). Distortion reduction in Slow Light systems based on stimulated Brillouin scattering, *Opt. Express*, Vol.16,No.11,pp. 8280-8285, eISSN1094-4087.

Herraez M. G, Song K. Y., and Thévenaz L. (2006), Arbitrary-bandwidth Brillouin slow light in optical fibers, *Opt. Express* , Vol.14, No.4, pp.1395-1400, eISSN1094-4087.

Zhu Z., Dawes A. Gauthier M. C., D. J., Zhang L., and Willner A. E. (2006), 12-GHz-Bandwidth SBS Slow Light in Optical Fibers, in *Proc. of OFC 2006*, paper PD1.

Song K. Y., Hotate K. (2007), 25 GHz bandwidth Brillouin slow light in optical fibers, *Opt. Lett.* , Vol.32 , No.3, pp.217-219, ISSN0146-9592.

Zhu Z., Gauthier D. J., and et al(2005), Numerical study of all-optical slow-light delays via stimulated Brillouin scattering in an optical fiber, *J. Opt. Soc. Am. B*, Vol.22 , No.11, pp.2378-2384, ISSN0740-3224.

Kalosha V. P., Chen L, and Bao X. Y(2006), Slow and fast light via SBS in optical fibers for short pulses and broadband pump, *Opt. Express*, Vol.14, No.26, pp. 12693-12703, eISSN1094-4087

Damzen M. J., Vlad V. I., Babin,V. and Mocofanescu A(2003), *Stimulated Brillouin Scattering: Fundamentals and Applications*, IOP Publishing, ISBN0 7503 0870 2.

Dane C. B., Neuman W. A. and Hackel L. A. (1994), High-energy SBS pulse compression, *IEEE J. Quantum Electro*, Vol.30, No.8, pp.1907-1915, ISSN0018-9197.

Zhu Z., Gauthier D. J., Okawachi Y, *et al*(2005). Numerical study of all-optical slow-light delays via stimulated Brillouin scattering in an optical fiber, *J. Opt. Soc. Am. B*, Vol.22, No.11, pp. 2378- 2384, ISSN0740-3224.

Velchev I., Neshev D., Hogervorst W., and Ubachs W(1999). Pulse compression to the subphonon lifetime region by half-cycle gain in transient stimulated Brillouin scattering. *IEEE J. Quantum Electron*, Vol.35, No.12, pp.1812-1816, ISSN0018-9197.

Part 2

Other Impairments in Optical Fibers

6

Optical Fiber Birefringence Effects – Sources, Utilization and Methods of Suppression

Petr Drexler and Pavel Fiala
Department of Theoretical and Experimental Engineering,
Brno University of Technology,
Czech Republic

1. Introduction

The application area of optical fibers is quite extensive. Telecommunication applications were the primary field of fibers employment. The related area is the utilization of optical fibers for control purposes, which benefits from principal galvanic isolation between the transmitting and receiving part of the system. A minimal sensitivity of light propagation inside the fiber to electromagnetic field of common magnitudes allows use the fiber in systems with high level of electromagnetic disturbance. Regarding the physical aspects of light propagation in fibers, they find utilization possibility in physical quantities sensors. It is possible to modulate the phase and state of polarization of the wave inside fiber optical medium by means of external physical quantity. The interaction is described by electro-optical, magnet-optical and elasto-optical effects.

In order to achieve high data transmission rates in field of telecommunication applications the single-mode fibers are used exclusively. Similarly, single-mode fibers are used in the case of intrinsic fiber optic sensors. Intrinsic fiber optic sensors exploit the fiber itself to external quantity sensing and the fiber serves to signal transmission also. The reason of single-mode fiber utilization is the presence of basic waveguide mode – single wave with single phase and single polarization characteristic.

In spite of the fiber utilization advantages we have to take into account undesirable effects, which are present in real non-ideal optical fiber. In telecommunication and sensor application field the presence of inherent and induced birefringence is crucial. The presence of birefringence may cause an undesirable state of polarization change. In the case of high-speed data transmission on long distances the polarization mode dispersion may occur. Due to this effect the light pulses are broadened. This may result in inter-symbol interference. In the case of sensor application, when the state of polarization is a carrier quantity, the possibility of output characteristic distortion and sensors sensitivity decreasing may occur.

It's advantageous to consider fiber sensor application for purposes of birefringence origin and influence description, since the presence of linear and circular birefringence together is watched often. While the inherent circular birefringence is negligible in common single-mode fibers, the inherent and induced linear birefringence may be present in considerable rate. The inherent linear birefringence is mostly undesirable effect, when we exclude utilization in polarization maintaining fibers. Whereas, the induced linear birefringence may be utilized for sensing purposes, e.g. for mechanical stress or pressure sensing. Similarly, induced circular birefringence is a principle effect for group of polarimetric sensors, e.g. polarimetric current sensor.

For the suppression of unwanted linear birefringence influence, inherent and induced also, several approaches and methods have been developed and published. They are often based on different principles. However, they differ in view of their properties and suitability for various applications.

The goal of this chapter is to present basic effects, which lead to occurrence of linear and circular birefringence in single-mode fibers. The methods, which may be used in order to suppress unwanted birefringence, will be presented also. Since the main manifestation of birefringence effects is the transformation of the polarization state of transmitted light, a brief recapitulation of basic polarization states and their illustrative visualizations are given. In following subchapters various mechanisms, which induce birefringence will be introduced together with corresponding relations and comprehensible illustrations. In the last descriptive subchapter the most significant methods for unwanted linear birefringence are presented with their properties and references to related literature.

2. Light polarization

The electromagnetic wave polarization represents how varies orientation, pertinently projection magnitude, of electric field component in a plane which is perpendicular to propagation direction. The polarization character of the wave may be described by means of the magnetic field component also. However, the interaction of matter with light wave is done mainly via the electric field component. Then the electric field intensity vector E is used for polarization states description usually. The general polarization state of the wave is the elliptical one. The special cases are the circular birefringence and linear birefringence.

Consider an electromagnetic wave which is described by electric field intensity vector E and which propagates in direction of z axis. The wave may be represented as a superposition of two partial waves with mutually orthogonal linear polarizations and with the same frequency

$$E = E_x + E_y,$$ (1)

where xyz is orthogonal coordinate system and E_x, E_y are vectors of electrical field intensity, which are aligned in x axis direction and y axis direction. It should be noted, that we consider the same frequency of both waves in all of the following analysis. In case of linear polarizations the electric field intensity vectors E_x, E_y swing along a straight line. These two vectors may be assigned to two degenerated modes of the single-mode fiber, which is a dielectric circular waveguide. In a lossless medium hold for field components magnitude relations

$$E_x(z,t) = E_{0,x} \cos(kz - \omega t + \phi_x),$$
$$E_y(z,t) = E_{0,y} \cos(kz - \omega t + \phi_y),$$ (2)

where $E_{0,x}$, $E_{0,y}$ are wave amplitudes, ω is angular frequency of the waves, t is time and ϕ_x, ϕ_y are phases of the wave, k is magnitude of the wave vector. Amplitudes ratio of $E_{0,x}$ and $E_{0,y}$ and phase difference $\Delta\phi = \phi_x - \phi_y$ determine the state of the polarization of the resulting wave.

In case when $E_{0,x} = E_{0,y}$ and $\Delta\phi = 0$ the orthogonal waves are in phase with the same amplitude. We obtain a linearly polarized wave by their superposition. Its plane of polarization is in 45° to y axis (or -45° to x axis) as shown in Fig. 1. In Fig. 1 and following figures k represents the wave vector.

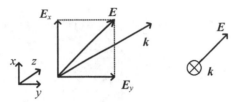

Fig. 1. Superposition of in-phase wave equal in amplitude results in linear polarized wave.

In case when $E_{0,x} = E_{0,y}$ and $\Delta\phi = \pm\pi/2$ the resultant wave has a circular polarization. The end point of E vector of circular polarized wave traces a circle. We differ between right-handed and left-handed circular polarized wave depending on the phase difference $\Delta\phi$ polarity, plus or minus. An illustration of right-handed circular polarized wave is shown in Fig. 2.

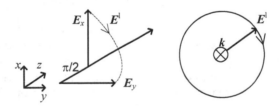

Fig. 2. Right-handed circular polarized wave.

When $\Delta\phi \neq 0, \pm\pi/2$ or $E_{0,x} \neq E_{0,y}$, we obtain an elliptically polarized wave, right-handed or left-handed, in dependence on phase difference $\Delta\phi$ polarity. The end point of E vector of elliptically polarized wave traces an ellipse. The case of left-handed elliptically polarized wave is shown in Fig. 3

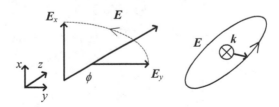

Fig. 3. Left-handed elliptically polarized wave.

As the light wave propagates in homogenous isotropic medium, its velocity remains constant independently on the propagation direction. The propagation velocity is given by the refractive index of the medium n. Refractive index is a ratio of wave velocity in vacuum c and wave phase velocity v_p in the medium, $n = c/v_p$. However, medium may be of anisotropic character. This means, that the propagation velocity depends on the propagation direction, pertinently polarization. This effect is observed in birefringent materials. In common birefringent materials the optical properties are described by means of index ellipsoid, which is shown in Fig. 4. When linearly polarized wave travels in z axis direction and it is polarized in y axis direction, the wave phase velocity is given by refractive index n_y. When the same wave will be polarized in x axis direction, the phase velocity will be given by refractive index n_x. Since $n_y > n_x$, the first wave will travel slower.

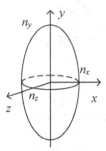

Fig. 4. Index ellipsoid example of birefringent material.

The velocity of wave polarized between y and x axis directions will be given by refractive index, which magnitude lies on ellipse in xy plane. When a light wave travels in birefringent medium of such type described above, it may occur a phase shift between its orthogonal components, which are described by relation (2). This occurs due to different propagation velocities of the components. The resulting state of polarization depends on total phase difference $\Delta\phi$, which is a function of propagation length in birefringent medium also. An example of state of polarization change from linear to elliptical by birefringent medium crossing is shown in Fig. 5.

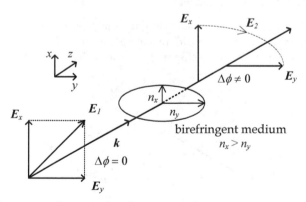

Fig. 5. State of polarization change in birefringent medium.

3. Linear birefringence in optical fiber

In previous section a polarization state of wave has been explained as a superposition result of two partial waves with certain phase shift and certain amplitude ratio. The similar concept may be used for description of polarization state transformation in single-mode optical fiber. As has been mentioned above, two degenerate modes HE_{11}^x and HE_{11}^y may exist in the fiber with circular core cross-section. The superposition of the two modes, which are orthogonal, results in the wave propagating in the fiber. And, the phase shift of these two modes determines the polarization state of the wave in the fiber. A deeper analysis of mode theory of fibers is out of the scope of this chapter and may be found in relevant literature (Iizuka, 2002).

The phase velocities of two orthogonal modes in the fiber $v_{f,x}$ and $v_{f,y}$ are given by magnitudes of wave numbers β_x and β_y of the modes

$$v_{f,x} = \frac{2\pi f}{\beta_x}, \tag{3}$$

$$v_{f,y} = \frac{2\pi f}{\beta_y}, \tag{4}$$

where f is the frequency of the wave. An ideal single-mode fiber with circular core cross-section along its length, made from homogenous isotropic material, will exhibit the same refractive index n for both of the modes. The wave numbers β_x and β_y will be equal also. The modes will propagate with the same phase speed v_f. At this condition the modes remain degenerate and the resulting polarization state will be preserved. A non-ideal fiber has not a constant circular core cross-section along its length or it exhibits anisotropy due to bending or other mechanical stress. As a consequence, the loss of modes degeneracy occurs. The fiber will behave as a birefringent medium with different refractive indices n_x and n_y and different phase velocities $v_{f,x}$ and $v_{f,y}$. In case of constant core cross-section and constant anisotropy, we can designate β_y as a wave number for a fast mode and β_x for a slow mode. Corresponding axes x and y may be designated as a fast axis and a slow axis of the fiber.

If a linear polarized wave is coupled into the birefringent fiber with gradually varying core cross-section or varying anisotropy, it is not possible to designate one mode as a fast one and second as a slow one. The mode phase shift $\Delta\phi$, which determines the output polarization state, is dependent on average wave number magnitudes and on the fiber length

$$\Delta\phi = \left(\overline{\beta_x} - \overline{\beta_y}\right)l_v. \tag{5}$$

The output polarization state will not be stable, when one would manipulate with the fiber or when the ambient temperature fluctuates. Since the wave number will be changing. This fact complicates the utilization of single-mode fibers in application with defined polarization state, as the fiber lasers or fiber sensors. Further, photodetectors, which are used in the field of fiber optic telecommunication, are not sensitive to polarization state. However, owing to the fiber birefringence, the phase shift of partial modes, pertaining to individual pulses, occurs. This effect causes a broadening of the impulses resulting in inter-symbol interferences.

The fiber birefringence rate is characterized by beat length l_b. It is possible to deduce from (5), that the state of polarization will transform periodically, as shows Fig. 6. Linear polarization of the wave with the polarization plane at angle $45°$ to x axis gradually transforms across right-handed elliptical polarization to right-handed circular polarization. It transforms further across right-handed elliptical and linear polarization perpendicular to the original one. Then it transforms across left-handed elliptical and left-handed circular to perpendicular left-handed elliptical polarization and finally to original linear polarization. At this point, the total phase shift of the modes is $\Delta\phi = 2\pi$ and the corresponding fiber length is the fiber beat length l_b.

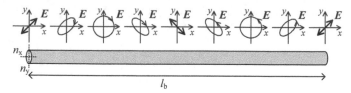

Fig. 6. Periodical transformation of the state of polarization in fiber with beat length l_b.

$$l_b = \frac{2\pi}{\beta_x - \beta_y} = \frac{\lambda}{n_{x,\text{eff}} - n_{y,\text{eff}}} = \frac{\lambda}{\Delta n_{\text{eff}}}, \tag{6}$$

where $n_{x,\text{eff}}$ and $n_{y,\text{eff}}$ are effective refractive indices in x axis and in y axis, Δn_{eff} is the difference of effective refractive indices and λ is light wavelength.

3.1 Linear birefringence owing to elliptical fiber core cross-section

As mentioned above, the linear birefringence may be of latent or induced nature. The main cause of latent linear birefringence in real fiber is the manufacture imperfection. The cross-section of the fiber core is not ideally circular but slightly elliptical, as shown in Fig. 7.

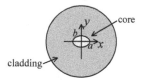

Fig. 7. Elliptical cross-section of the non-ideal fiber core.

Let the major axis of the ellipse representing core cross-section lies in the x axis direction and the minor axis lies in the y axis direction. The wave number of the mode, which propagates in x axis direction, will be of a larger magnitude than the wave number of the y axis mode. The difference of effective refractive indices Δn_{eff} is determined by the ellipticity ratio a/b. In case of small ellipticity rate, when $a \approx b$, holds the relation

$$\Delta n_{\text{eff}} = 0.2 \left(\frac{a}{b} - 1 \right) (\Delta n)^2, \tag{7}$$

where $\Delta n = n_{co} - n_{cl}$ is the difference of core refractive index n_{co} and cladding refractive index n_{cl}. For specific phase shift of the modes it may be derived from (7) a relation

$$\Delta\phi = \frac{0.4\pi}{\lambda} \left(\frac{a}{b} - 1 \right) (\Delta n)^2. \tag{8}$$

In order to attain a large beat length the fiber core should approach the circular cross-section as much as possible. It may be derived by means of (6) and (8) a demand for relative deviation from ideal circularity

$$\left(\frac{a}{b}-1\right)\cdot 100\% = \frac{\lambda}{0.2l_b\left(n_{co}-n_{cl}\right)^2}\cdot 100. \tag{9}$$

For typical single-mode fiber with $n_c = 1.48$, $n_{cl} = 1.46$ and operating wavelength $\lambda = 633$ nm the required deviation from ideal circularity achieves 0.016%. This demand is very hard to accomplish in fiber manufacture. The common fibers maintain the polarization state close to the initial only a few meters along.

3.2 Inner mechanical stress induced linear birefringence

The fiber core ellipticity is not a single source of fiber birefringence imposed by the manufacture. A second important source, which may take effect, is the presence of inner mechanical stress on the core. This may be caused by non-homogeneity of cladding density in area close to the core. In order to simplify the analysis we can consider an elliptical density distribution owing to imperfect technology process of fiber drawing from hot preform. The far area of the cladding influences the inner area by centripetal pressure after the fiber cooled down. Since the core-close area has a non-homogenous density, the pressures on core, p_x and p_y, will act non-uniformly as illustrates Fig. 8.

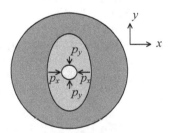

Fig. 8. Non-uniform stress on fiber core owing to imperfect inner structure.

Due to the photo-elastic effect, which causes pressure dependent anisotropy, the fiber core becomes a birefringent medium. Then, the difference of effective refractive indices in x axis and y axis is

$$\Delta n_{eff} = \frac{C_f}{1-v_c}\Delta v\Delta T\frac{p_x/p_y-1}{p_x/p_y-1}, \tag{10}$$

where v_c is Poisson constant of the core, Δv is difference of expansion coefficients of outer and inner cladding areas, ΔT is difference between softening temperature of the cladding and the ambient temperature. Coefficient C_f is characteristic for given fiber, given by

$$C_f = \frac{1}{2}\left(\frac{n_{co}-n_{cl}}{2}\right)^3\left(r_{11}-r_{12}\right)\left(1-v_{co}\right), \tag{11}$$

where r_{11}, r_{12} are components of photo-elastic tensor matrix of the fiber material. Photo-elastic matrix description is above the scope of this chapter and may be found in (Huard, 1997). For specific phase shift of the modes it may be derived from (10) a relation

$$\Delta\phi = \frac{2\pi}{\lambda}\frac{C_f}{1-v_c}\Delta v\Delta T\frac{\dfrac{p_x}{p_y}-1}{\dfrac{p_x}{p_y}-1}.\tag{12}$$

The influence of inner stress induced linear birefringence is weak in compare to birefringence owing to elliptical core cross-section in common single-mode fibers. However, the inner stress induced linear birefringence may be imposed intentionally in case of polarization maintaining (PM) fibers manufacture.

3.3 Outer mechanical stress induced linear birefringence
Linear birefringence in single mode fiber may be induced by outer influence also. It is caused by outer mechanical stress (pressure or tensile force) on fiber cladding. Cladding transfers the mechanical stress on the core and similar effect described above uprises. In practice, an action of force in one dominant direction appears usually. It induces origin of two axis of symmetry, x and y, with two refractive indices n_x and n_y again.

One of the possible effects causing linear birefringence is fiber bending, which is illustrated in Fig. 9. A fiber with cladding diameter d_{cl} is bended with diameter R. The fiber axis is equal to y axis direction. The pressure imposed on core in x axis increase refractive index n_x in compare to n_y due to the photo-elastic effect. In this configuration, the slow mode propagates in the bending plane xy and the fast mode propagates in plane yz.

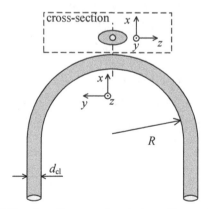

Fig. 9. Geometric relation of fiber bending causing induced linear birefringence.

The difference of effective refractive indices in x axis and y axis will be (Huard, 1997)

$$\Delta n_{\text{eff}} = \frac{1}{2}C_f\frac{d_{cl}^2}{R^2}+2\zeta C_f\frac{d_{cl}}{R}\tag{13}$$

and related specific phase shift of the modes is expressed as

$$\Delta\phi = \frac{2\pi}{\lambda}\left[\frac{1}{2}C_f\frac{d_{cl}^2}{R^2} + 2\zeta C_f\frac{d_{cl}}{R}\right], \tag{14}$$

where C_f is fiber coefficient given by (11), ζ is the rate of axis deformation caused by longitudinal tensile force. The terms on the right side of relations (13) and (14) represent a situation when an additive tensile force acts on the fiber. This may occur when the fiber is bended over a solid, as a coil core. If the fiber is bended without the additive tensile force relations (13) and (14) are simplified. Then, the resultant relation for specific phase shift is

$$\Delta\phi = \frac{\pi}{\lambda}C_f\frac{d_{cl}^2}{R^2}. \tag{15}$$

Relation (15) may be substituted by relation, where the fiber coefficient C_f is replaced by the product of Young module of the fiber material E_c and photo-elastic coefficient of the fiber core $\Re$ (Ulrich et al., 1980)

$$\Delta\phi = \frac{\pi}{\lambda}E_c\Re\frac{d_{cl}^2}{R^2}. \tag{16}$$

As in previous cases, it may be derived from (16) a relation for specific phase shift of the modes

$$\Delta\phi_{1z} = \frac{2\pi^2}{\lambda}E_c\Re\frac{d_{cl}^2}{R}. \tag{17}$$

When a one fiber turn with radius $R = 8$ cm would be formed from a typical single-mode fiber with $Ec = 7.45 \cdot 10^9$ Pa, $\Re = -3.34 \cdot 10^{-11}$ Pa^{-1} (Namihira, 1983) and $d_{cl} = 125$ μm, the phase shift at wavelength $\lambda = 633$ nm achieves $\Delta\phi \approx -\pi/2$. In this case, for example linear polarization will be transformed into the circular. The original polarization state will be lost. A second significant effect, which induces linear birefringence in the fiber is a lateral pressure, which is illustrated in Fig. 10.

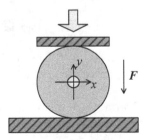

Fig. 10. Imposing a lateral pressure force on the fiber.

Induced anisotropy in the fiber is a result of photo-elastic effect, which is induced by compressing fiber between two planar solid slabs. If we consider F_m as a force acting on unit length, the phase difference of the fiber modes will be (Huard, 1997)

$$\Delta\phi = \frac{2\pi}{\lambda}C_f\frac{4F_m}{\pi d_{cl}E_c}. \tag{18}$$

The fast mode will propagate along the x axis and the slow one along the y axis. Lateral pressure induced birefringence may occur by fiber assembly in to optical components, such as connectors.

At the close of this chapter, there should be mentioned another way to induce the fiber linear birefringence. It may be imposed by electro-optical effect in the fiber core. However, the fiber core is made from amorphous material and the electro-optic effect is of a very weak character (Wagner et al., 1992).

4. Circular birefringence in optical fiber

In the case of circular birefringence analysis we introduce a concept of chiral birefringent medium. It exhibits two refractive indices n^r and n^l for right-handed and left-handed circular polarized waves. Counter rotating waves, which propagate in this medium, travel with different phase velocities and they gain a phase shift. Both of the circular polarized waves we may decompose on two linear polarized waves with equal amplitudes and with $\pi/2$ or $-\pi/2$ phase shift. Thus, the right-handed circular polarized wave propagating in z axis direction is a superposition of two orthogonal linear polarized waves described by components E_x^r and E_y^r. For their magnitudes holds

$$E_x^r(z,t) = \frac{E_0}{2}\cos\left(\beta^r z - \omega t - \frac{\pi}{2}\right),$$

$$E_y^r(z,t) = \frac{E_0}{2}\cos\left(\beta^r z - \omega t\right),$$

(19)

where β^r is wave number of right-handed circular polarized wave. Likewise, for left-handed circular polarized wave components holds

$$E_x^l(z,t) = \frac{E_0}{2}\cos\left(\beta^l z - \omega t + \frac{\pi}{2}\right),$$

$$E_y^l(z,t) = \frac{E_0}{2}\cos\left(\beta^l z - \omega t\right),$$

(20)

where β^l is wave number of left-handed circular polarized wave. Wave numbers of waves propagating in fiber core are given as

$$\beta^r = \frac{2\pi}{\lambda}n_j^r, \ \beta^l = \frac{2\pi}{\lambda}n_j^l.$$

(21)

Due to the magnitude difference of refractive indices n_j^r and n_j^l in circular birefringent fiber core, the counter rotating waves travel with different phase velocities and they gain a phase shift

$$\Delta\phi = \left(\beta^r - \beta^l\right)l_f = \frac{2\pi}{\lambda}\left(\beta_j^r - \beta_j^l\right)l_f = \sigma_c l_f,$$

(22)

where σ_c is specific rotation of fiber core and l_f is fiber length.

When we superpose two circular polarized waves, which were described above, we obtain a linear polarized wave with a certain orientation of polarization plane. The change of polarization plane rotation angle $\Delta\alpha$ is equal to the phase shift $\Delta\phi$ from (22), $\Delta\alpha = \Delta\phi$.

It may be concluded, that the presence of circular birefringence in the fiber results in polarization plane rotation. When the fiber is free from linear birefringence and we couple a linear polarized wave into the fiber, we obtain a linear polarized wave with rotated polarization plane at the output. The angle of plane rotation is due to the circular birefringence rate and the fiber length.

In contrast to linear birefringence, circular birefringence of latent origin is negligible in common single-mode fiber. Nevertheless, it is possible to impose it in manufacturing process or induced it by outer influence. This can be attained by suitable applied mechanical stress or by magnetic field applying in the direction of fiber axis.

4.1 Outer mechanical stress induced circular birefringence

If a fiber section with length l_f is exposed to torsion with specific torsion rate τ

$$\tau = \frac{\delta}{l_f}, \tag{23}$$

where δ is torsion angle as shown in Fig. 11, a sheer stress is imposed in plane perpendicular to fiber axis.

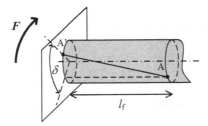

Fig. 11. A fiber section with length l_f exposed to torsion with angle δ.

Imposed sheer stress results in fiber core anisotropy owing to photo-elastic effect. In order to describe optical properties of anisotropic fiber core, it is useful to exploit tensor matrix of dielectric constant ε (Saleh & Teich, 1991)

$$\varepsilon = \varepsilon_i + \Delta\varepsilon_t = \begin{bmatrix} \varepsilon & 0 & 0 \\ 0 & \varepsilon & 0 \\ 0 & 0 & \varepsilon \end{bmatrix} + \begin{bmatrix} 0 & -g\tau y & 0 \\ g\tau y & 0 & -g\tau x \\ 0 & g\tau x & 0 \end{bmatrix} = \begin{bmatrix} \varepsilon & -g\tau y & 0 \\ g\tau y & \varepsilon & -g\tau x \\ 0 & g\tau x & \varepsilon \end{bmatrix}, \tag{24}$$

where ε_i is dielectric constant tensor of original medium and $\Delta\varepsilon_t$ is tensor of torsion contribution. Coordinates x,y in matrix $\Delta\varepsilon_t$ belongs to point A' in sheer stress plane, where is the dielectric constant expressed and for coefficient g holds

$$g = r_{44}n_c^4 = (r_{11} - r_{12})n_c^4, \tag{25}$$

where r_{11}, r_{12} and r_{44} are components of photo-elastic matrix of the fiber core material. For further analysis, it is advantageous to exploit a Jones calculus (Jones, 1941) to characterize the influence of torsion modified medium on the polarization state of the wave. The relations of photo-elastic coefficients of the medium and Jones matrix of the medium are beyond the scope of this chapter and may be found for example in (Iizuka, 2002). The Jones matrix of the torsion modified medium is in the form

$$T_c = \begin{bmatrix} 0 & -jg\tau \\ jg\tau & 0 \end{bmatrix}, \qquad (26)$$

where j is imaginary unit. Jones matrix T_c describes the polarizing properties of circular birefringent medium. If we multiply matrix T_c with Jones vector J_1 of linear polarized wave, we obtain vector J_2 with imaginary components. Both of the components represent left-handed and right-handed circular polarized waves.

$$J_2 = T_c \cdot J_1 = \frac{1}{\sqrt{2}} \begin{bmatrix} 0 & -jg\tau \\ jg\tau & 0 \end{bmatrix} \cdot \begin{bmatrix} 1 \\ 1 \end{bmatrix} = \frac{1}{\sqrt{2}} \begin{bmatrix} -jg\tau \\ jg\tau \end{bmatrix}. \qquad (27)$$

The phase shift of circular polarized waves $\Delta\phi$ and corresponding polarization rotation angle $\Delta\alpha$ is proportional to the torsion rate τ. Then, a twisted single mode fiber with length l_f acts as a polarization rotator with rotation angle

$$\Delta\alpha = g\tau l_f. \qquad (28)$$

4.2 Magnetic field induced circular birefringence

The second source of fiber circular birefringence is magneto-optical effect. Between three types of magneto-optical effect (Cotton-Mouton, Kerr, Faraday) (Craig & Chang, 2003), the Faraday effect is significant for silica fiber. It induces circular birefringence owing to magnetic field action in direction along the fiber axis. Analogous to fiber torsion, the Faraday magneto-optical effect modifies the dielectric constant tensor

$$\varepsilon = \varepsilon_i + \Delta\varepsilon_{mo} = \begin{bmatrix} \varepsilon & 0 & 0 \\ 0 & \varepsilon & 0 \\ 0 & 0 & \varepsilon \end{bmatrix} + \begin{bmatrix} 0 & -j\eta B & 0 \\ j\eta B & 0 & 0 \\ 0 & 0 & 0 \end{bmatrix} = \begin{bmatrix} \varepsilon & -j\eta B & 0 \\ j\eta B & \varepsilon & 0 \\ 0 & 0 & \varepsilon \end{bmatrix}, \qquad (29)$$

where $\Delta\varepsilon_{mo}$ is tensor of magneto-optical effect contribution, B is the magnitude of flux density of the external magnetic field, η is coefficient, which is proportional to magneto-optic specific rotation coefficient (Huard, 1997). Again, dielectric constant tensor (29) describes a birefringent medium, where right-handed and left-handed circular polarized waves travel with different velocities. Here, the resulting phase shift of the waves is proportional to magnitude of magnetic flux density and the length of birefringent medium. In order to explain the origin of Faraday magneto-optical effect, it is possible to model the effect as an electron oscillator movement in magnetic field (Waynant & Ediger, 2000). The effect itself results from interaction of outer magnetic field with oscillating electron, which is excited by the electric field of the light wave. Electrons represent harmonic oscillators. For them equations of forced oscillations hold. In the presence of external magnetic field with flux density B, parallel to wave propagation direction, for the electron oscillator holds

$$m_e \frac{d^2 u}{dt^2} + \kappa u = -eE - e\left[\frac{du}{dt} \times B\right], \qquad (30)$$

where m_e is electron mass, e is electron charge, u is vector, which determines the electron displacement, κu is quasi-elastic force preserving electron in equilibrium position, E is electric field vector of propagating wave. Electric field of the wave polarizes the medium

$$P = -N_e e u, \tag{31}$$

where N_e is the count of electrons in volume unit, which are deflected by the electric field of the wave. Substituting equation (31) into (30) we get

$$\frac{d^2 P}{dt^2} + \frac{e}{m_e}\left[\frac{dP}{dt} \times B\right] + \omega_0^2 P = \frac{N_e e^2}{m_e} E, \tag{32}$$

where ω_0 is frequency of the electron oscillator. Equation (32) represents the system of two simultaneous differential equations. We obtain two terms by their solution. One for the right-handed, second for the left-handed circular polarized wave in the medium (Born & Wolf, 1999)

$$E^\mathbf{r} = E_0^\mathbf{r} e^{j\omega t},$$
$$E^\mathbf{l} = E_0^\mathbf{l} e^{j\omega t}, \tag{33}$$

where ω is frequency of circular polarized waves. The macroscopic relation for the medium polarization due to the electric field of circular polarized waves is in the form

$$P^\mathbf{r} = \varepsilon_0 \chi^\mathbf{r} E^\mathbf{r},$$
$$P^\mathbf{l} = \varepsilon_0 \chi^\mathbf{l} E^\mathbf{l}, \tag{34}$$

where $\chi^\mathbf{r}$ and $\chi^\mathbf{l}$ are dielectric susceptibilities for right-handed and left-handed circular polarized waves and ε_0 is dielectric constant of vacuum. Refractive index of the medium is related to dielectric susceptibility

$$n^2 = \varepsilon_r = 1 + \chi, \tag{35}$$

where ε_r is relative dielectric constant of the medium. Substituting equations (34) into system (32) and by utilization of relation (35), we obtain relations for refractive indices of right-handed and left-handed circular polarized waves

$$\left(n_c^\mathbf{r}\right)^2 = 1 + \frac{N_e e^2}{\varepsilon_0 m_e} \cdot \frac{1}{\omega_0^2 - \omega^2 + \dfrac{e}{m_e}B\omega},$$

$$\left(n_c^\mathbf{l}\right)^2 = 1 + \frac{N_e e^2}{\varepsilon_0 m_e} \cdot \frac{1}{\omega_0^2 - \omega^2 - \dfrac{e}{m_e}B\omega}. \tag{36}$$

When we take into account certain simplifications, we can differentiate equations (36) and we can derive relation for polarization plane rotation in dependence on the outer magnetic field flux density B and on the interaction length l_f (fiber length in magnetic field)

$$\Delta\alpha = \Delta\phi = \frac{\mathrm{\pi}}{\varepsilon_0\lambda}\frac{N_e}{n_c}\frac{e^3}{m_e^2}\frac{\omega}{\left(\omega_0^2 - \omega^2\right)^2}Bl_f = VBl_f,$$
(37)

where λ is wavelength of the wave, $\overline{n_c} = (n_c{}^r + n_c{}^l)/2$ is the mean refractive index, ω is angular frequency of the wave, V is Verdet constant, which characterizes magneto-optic properties of medium. It is obvious, that Verdet constant depends on the wavelength.

The right part of equation (37) is the basic relation for Faraday magneto-optic effect. The effect is non-reciprocal. The polarization rotation direction depends on the mutual orientation of magnetic flux density B and the wave propagation direction. The polarization of wave propagating in the direction of B experiences a rotation $\Delta\alpha$. The polarization of wave propagating in the opposite direction to B experiences a rotation $-(\Delta\alpha)$. This non-reciprocal character is important for example for polarization mode conjugation as will be shown later. The illustration of polarization plane rotation in fiber section due to Faraday magneto-optic effect is shown in Fig. 12.

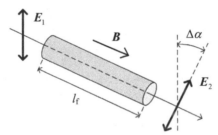

Fig. 12. Polarization plane rotation in fiber section due to Faraday magneto-optic effect.

5. Superposition of linear and circular birefringence in fiber

Both types of birefringence, linear and circular, may appear in single mode fiber. Both of them may be of latent or induced origin. The total phase shift of modes in fiber, determining the output polarization, is given by their geometrical average

$$\Delta\phi = \sqrt{\phi_c^2 + \left(\frac{\phi_l}{2}\right)^2},$$
(38)

where ϕ_c is mode phase shift caused by circular birefringence and ϕ_l is mode phase shift caused by linear birefringence (Ripka, 2001). Generally, the polarization state of the output wave will be elliptical due to the linear birefringence and the orientation of axes of the polarization ellipse will rotate due to the circular birefringence. The analysis of the optical system with fiber, which exhibits both types of birefringence, may be performed by means of Jones calculus. Jones matrix of the fiber will be in the form (Tabor & Chen, 1968)

$$T_f = \begin{bmatrix} \cos\Delta\phi + j\dfrac{\phi_l}{2}\dfrac{\sin\Delta\phi}{\Delta\phi} & -\phi_c\dfrac{\sin\Delta\phi}{\Delta\phi} \\[3mm] \phi_c\dfrac{\sin\Delta\phi}{\Delta\phi} & \cos\Delta\phi - j\dfrac{\phi_l}{2}\dfrac{\sin\Delta\phi}{\Delta\phi} \end{bmatrix},$$
(39)

where $\Delta\phi$ results from (38). By means of (39), it is possible to study transformation of polarization state of the wave, which passed trough the fiber. Generally, in presence of both types of birefringence, the fiber behaves as phase retarder and polarization rotator simultaneously.

6. Techniques for unwanted fiber birefringence suppression

In previous chapters, it has been explained how the birefringence affects the polarization state of the wave in fiber. The transformation of polarization state is often unwanted, if we intend to use it as a carrier quantity.

It is important to suppress the polarization mode dispersion in telecommunication applications, in order to avoid pulses broadening. It is caused by linear birefringence. Important is also to avoid the unwanted birefringence in polarimetric sensors applications. It has to be ensured, that the polarization state will be modified by sensing quantity only. Polarimetric fiber optic sensor may be divided into two groups. Sensors of mechanical quantities (strain, pressure, vibrations) utilize induced linear birefringence. Sensors of magnetic field utilize induced circular birefringence. Since the inherent circular birefringence of common fibers is insignificant, the key parameter is the rate of linear birefringence.

The facts mentioned above place demands for methods for unwanted linear birefringence suppression. Following subchapters present a brief overview of the most significant selected methods, which are used to meet this requirement. The methods differ in view of its principle, efficiency or usability in various applications.

6.1 Polarization maintaining fibers

Polarization maintaining (PM) fibers have a specific inner structure, which allows maintaining polarization of the wave on long distances. In general view, polarization maintaining fibers may be divided in two groups. The first represents polarization maintaining fibers with low birefringence (PM LB). PM LB fiber approaches the concept of ideal fiber with constant circular cross-section and with very low linear birefringence. As has been mentioned above, these fibers are difficult to manufacture. Moreover, the manipulation (as bending or compressing) with fiber induces linear birefringence due to photo-elastic effect. In the second group belong polarization maintaining fibers with high birefringence (PM HB). A strong linear birefringence is imposed in the fiber by means of internal mechanical strain, which results in the loss of degeneracy of hybrid fiber modes HE_{11}. Therefore, the beat length of PM HB fibers is only a few millimeters. Hybrid modes propagate in fiber along the major and minor axis of ellipse, whose ellipticity is given by the ratio of mode wave numbers β_x and β_y. If the light wave, for example with linear polarization, is coupled into the fiber with polarization plane in direction of one of the axes, the both orthogonal wave modes will experience equal wave numbers and the equal refractive indices. The wave will propagate along the fiber without the polarization state transformation (Kaminow & Ramaswamy, 1979). The sensitivity on bending and temperature fluctuations is greatly reduced. Nevertheless, the insensitive polarization state preserving is ensured only for one certain polarization plane orientation, when both wave modes experience same refractive indices.

The principle of PM HB fibers manufacturing consists in implementing of stress components in the fiber cladding. Stress components impose symmetrical defined pressure force on circular fiber core. Stress components are implemented by doping of designated cladding areas with atoms of certain elements, typically boron atoms. In this way, areas with different thermal expansion coefficient are formed. After the drawn fiber cools down, the doped areas cause inner strain, which acts on the fiber core. Doped areas may have variety of shapes as shown in Fig. 13. The influence of fiber latent linear birefringence is strongly exceeded by the imposed birefringence, which temperature and bending dependence is very weak.

In order to characterize the properties of PM HB fibers the polarization crosstalk CT is defined (Noda et al., 1986). Polarization crosstalk, defined by relation (40), is given by logarithmic ratio of optical power of excited mode P_x and optical power of coupled mode P_y. The polarization crosstalk is typically lower than -40 dB for fiber length of 100 meters (Senior, 2009).

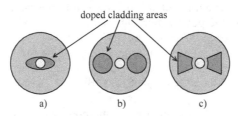

Fig. 13. Various profiles of PM HB fibers: a) elliptical, b) PANDA type, c) Bow-Tie type.

$$CT = 10\log\frac{P_y}{P_x}. \tag{40}$$

In the following chapters, fibers with intended polarization preserving properties are discussed. Although they exploit different principles, they may be considered as a special type of PM LB or PM HB fibers, as will be mentioned.

6.2 Fibers with high circular birefringence
The demands of telecommunication applications for polarization state preserving in fiber may be satisfactory covered by PM fibers. Since, PM fiber allows preservation of state polarization only for one certain polarization plane orientation, they are not well suited for applications in polarimetric fiber sensors. The polarization plane rotates due to the sensing quantity magnitude in case of these sensors. Therefore, the different fiber modifications were studied.

6.2.1 Twisted fibers
One of the approaches is fiber twisting, which may impose a strong circular birefringence in the core. If the rate of induced circular birefringence will be much greater than the rate of linear birefringence, relation (38) may be modified

$$\Delta\phi = \sqrt{\phi_c^2 + \left(\frac{\phi_l}{2}\right)^2} \overset{\phi_c \gg \phi_l}{\cong} \sqrt{\phi_c^2} = \phi_c. \tag{41}$$

The influence of circular birefringence will dominate and the linear birefringence may be neglected. Since the effect of linear birefringence is canceled, twisted fibers belong to group of PM LB fibers. Twisted fiber will behave as polarization plane rotator, which preserves the polarization state of the wave. The polarization plane rotation is proportional to specific rotation of the fiber core and σ_c and l_f is fiber length

$$\Delta\alpha = \sigma_c l_f. \tag{42}$$

If this principle would be utilized in polarimetric fiber sensor application, the polarization rotation due to sensed physical quantity will be additive to inherent polarization rotation of the fiber.

The circular birefringence in fiber core is possible to impose by fiber twisting in a plane, which is perpendicular to fiber longitudinal axis. The rate of circular birefringence corresponds to photo-elastic properties of the core, core refractive index, fiber length and specific torsion rate (relation (23)). In order to minimize the influence of linear birefringence, the specific torsion rate should be maximized. However, this is limited by torsion limit of the fiber. When exceeds, the fiber may be broken.

The disadvantage of the twisted fiber utilization in fiber sensing applications is the temperature dependence of twisting imposed circular birefringence, due to temperature dependence of core anisotropy. The next issue is the fabrication difficulty of small fiber coils for magnetic field sensing around conductors. The bending induced linear birefringence achieves a higher magnitude for small fiber coils. Therefore, the large torsion rate of the fiber has to be used and it may exceed the torsion limit. The approximate torsion limit of common single mode fiber is 100 turns per 1 meter (Payne et al., 1982). Taking into account this limit, it is possible to fabricate fiber coils with minimal diameter of 15 cm (Laming & Payne, 1989).

6.2.2 Spun low- and high-birefringent fibers

The more sophisticated approach to linear birefringence suppression was development of spun fibers. Spun fiber fabrication consists in twisting of melted preform during fiber drawing. During the fiber drawing, all fiber imperfections, as deviation from circularity and other non-uniformities, are spread out in all directions. Therefore, phase retardations, which experience propagating modes, cancel each other out. Since the twisted preform is melted, no stress induced anisotropy is present in the core after the fiber cools down. Simultaneously, the fiber is free from temperature dependency effects. Hence, the spun fiber behaves as an ideal fiber with circular cross-section core, which retain any polarization state of coupled wave, from the input to the output. In principle, spun fibers belong to group of PM LB fibers. We designate them as low-birefringent spun fibers (spun LB). Spun LB fiber exhibits only a negligible latent circular birefringence, due to limited viscosity of the preform during the drawing. The principle of spun LB fiber implies their main disadvantage, which is the sensitivity on fiber bending. This results in stress induced linear birefringence and the limitation for small radius fiber coil fabrication remains (Payne et al., 1982).

A similar concept of spun LB fiber represents highly birefringent spun fibers (Spun HB). Spun HB fibers are manufactured by rotating of melted preform also. However, the preform is prepared as for classical PM HB fiber, e.g. Bow-Tie (Laming & Payne, 1989). Spun HB fiber transforms the input linear polarization on to the elliptical. By carefully chosen rotation rate of the preform in relation to the fiber linear birefringence rate, it is possible to attain quasi-circular birefringence with negligible residual linear birefringence (Payne et al., 1982). Generally, we may consider spun HB fibers as a type of PM HB fiber group. The advantage of spun HB fibers is considerable immunity to rising of linear birefringence by fiber bending or compressing. Since the quasi-circular birefringence in the fiber originates from twisting of stress components, the temperature dependence of anisotropy is indispensable. Therefore temperature compensation has to be used in applications utilizing spun HB fibers. On the present, spun HB fibers for telecommunication and sensing application are available for wavelengths from 600 nm to 1600 nm, with attenuation in order of ones of dB·km-1. They may be wind on fiber coils with radius above 20 mm.

In connection with recent advances in microstructured fibers research, new possibilities of spun fiber fabrication emerge. The concept of microstructured fiber allows designing and producing of fibers with specific parameters on selected wavelength, single mode or multi mode character, polarization transformation properties and others. A development of microstructured spun fiber with six air chambers around the core and attenuation below 5 dB·km-1 is reported in (Nikitov et al., 2009). The possibility of circular polarization transmission has been achieved by rotating of the microstructured preform, together with magneto-optic properties preservation. The fiber coils with diameter above 2.5 mm can be fabricated for current sensing application.

6.3 Annealed fibers

Drawback of twisted fibers and spun HB fibers, which is anisotropy temperature dependence, limits their applicability mainly in polarimetric current sensor applications. A method for suppression of temperature dependence of anisotropy together with the suppression of bending induced linear birefringence has been proposed and experimentally studied (Stone, 1988; Rose et al., 1996). The method utilizes annealing of fabricated fiber coil. The procedure consists in temperature treatment of the fiber coil, which is installed in ceramic labyrinth. The coil is then heated up with approximate temperature-time gradient $\Delta T / \Delta t$ = 8·10-2 °C·s-1. When 850 °C is reached, the temperature is maintained for roughly 24 hours. Then, the slow cooling follows with approximate gradient $\Delta T / \Delta t$ = -3·10-3 °C·s-1. Annealed fiber coil is then transferred into protective case, which is filled with low-viscosity gel in order to damp the vibrations. The annealing procedure leads to removing of bending induced stress on the fiber and the linear birefringence is greatly suppressed. Prior to the annealing procedure, the fiber jacket and buffer has to be removed, because its oxidation at the temperatures 500 - 600 °C would damage the fiber. Since the fiber jacket and buffer act as fiber strength element, their removal is difficult and fiber rupture impends. The outer layers removal is facilitated by etching in organic solvent. The oxidation proceeds then without the negative influence on the fiber cladding (Rose et al., 1996).

Due to considerable temperature stability, the annealing method is used for fabrication of fiber current sensors, which are installed in outdoor environment on high voltage systems. The need for reliable galvanic isolation and accuracy predominates the technological difficulties in fiber coil fabrication. The annealing procedure has to be carefully performed and it has to be handled a technology of fiber coil isolation from vibrations. Since the fiber strength outer layers have been removed, the fiber coils have increased sensitivity to vibrations. For outdoor installations on high voltage system an annealed fiber sensors with sensitivity variation smaller than 0.2 %, dynamic range of 80 dB and temperature range 20-80 °C were developed (Higuera-Lopez, 2002).

6.4 Reciprocal compensation of linear birefringence

In combination with fibers, which were described above and with common single mode fibers also, another perspective approach for linear birefringence suppression may be exploited. The approach is based on reciprocity of linear birefringence. Beyond this, the circular birefringence is of non-reciprocal character, which gives usability especially in polarimetric current fiber sensors. Utilizing this fact, a compensation of linear birefringence may be performed on sensor output signal level in case of counter-propagating of two light waves in fiber. Second possibility consists in compensation of modes phase shift on fiber level in case of back-propagating of light wave with ortho-conjugated polarization.

6.4.1 Compensation on sensor output signal level
Since the linear birefringence is of reciprocal character, its influence on polarization state of the wave in fiber is not dependent on the propagation direction. The wave will experience the same polarization state transformation with the same orientation, no matter the propagation direction. Conversely, the magnetic field induced circular birefringence is non-reciprocal. When the wave will propagate in one direction, it will gain a polarization rotation. When the wave will propagate in opposite direction, it will gain rotation in opposite direction. The total rotation will be the double of the rotation in one direction.

Setups, which exploit the reciprocity of linear birefringence, have been demonstrated as polarimetric current sensors. An example of the sensor setup is shown in Fig. 14 (Claus & Fang, 1996).

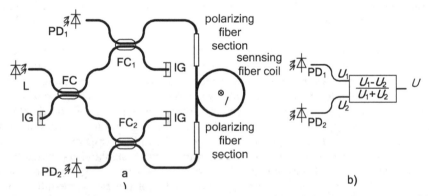

Fig. 14. Sensor setup with compensation on output signal level a), signal processing b).

In Fig. 14a), the signal from laser source L is divided into two channels by means of fiber coupler FC. After passing FC_1 and FC_2, the two optical signals propagate in opposite direction through the polarizing parts and the sensing part of the fiber. The unused coupler outputs are led into immersion gel to avoid reflections. The optical signals are sensed by photodetectors PD_1, PD_2. For output voltage signals of the detectors we can deduce

$$U_1 = R_{U,1} \cdot FC_1 \cdot T_{ov}^- \cdot FC_2 \cdot FC \cdot P_o,$$
$$U_2 = R_{U,2} \cdot FC_2 \cdot T_{ov}^+ \cdot FC_1 \cdot FC \cdot P_o,$$

$$(43)$$

where P_o is the input optical power. FC, FC_1, FC_2 are split ratio of coupler FC, FC_1 FC_2, T_{ov}^+ and T_{ov}^- are polarizing transfer functions of the fiber path in one and in second direction, $R_{U,1}$, $R_{U,2}$ are responsivities of photodetectors. For correct operation, it must hold $FC = FC_1 = FC_2 = 0{,}5$ and $R_{U,1} = R_{U,2} = R_U$. Relation (43) transforms into

$$U_1 = R_U \cdot FC^3 \cdot T_{ov}^-,$$
$$U_2 = R_U \cdot FC^3 \cdot T_{ov}^+.$$

(44)

In signal processing block, as shown in Fig. 14b), the normalized difference is computed

$$U = \frac{U_1 - U_2}{U_1 + U_2} = \frac{R_U \cdot FC^3 \cdot \left(T_{ov}^- - T_{ov}^+\right)}{R_U \cdot FC^3 \cdot \left(T_{ov}^- + T_{ov}^+\right)} = \frac{T_{ov}^- - T_{ov}^+}{T_{ov}^- + T_{ov}^+}.$$

(45)

Considering the presence of reciprocal linear birefringence only, the equality $T_{ov}^+ = T_{ov}^-$ holds. The output signal according to (45) will be zero. Since the magnetic field induced circular birefringence is non-reciprocal, the equality of polarizing transfer functions will not hold. Hence, the system will be responsive on varying rotation due to induced circular birefringence. Sensor utilizing above described principle dispose of considerable temperature stability and vibration insensitivity. More detailed description and sensor properties may be found in (Fang et al., 1994; Wilsch et al., 1996).

6.4.2 Orthogonal conjugation compensation
The reciprocity of undesirable birefringences in optical fiber may be used for their compensation exploiting the polarization ortho-conjugation of the wave modes. The method involves the back-propagation of light wave with conjugated modes through the same section of birefringent fiber.

As it has been stated above, imposing stress on fiber (pressure, bending) leads to origin of linear birefringent fiber core with two refractive indices, one lying in x axis direction – n_x and second lying in y axis direction – n_y. We may designate the axis as the fast fiber axis, with lower refractive index, and the slow fiber axis, with higher refractive index. However, the orientation of the fast and slow axes system towards the geometrical coordinate system of the fiber changes along, in dependence on the bending or pressure force orientation. The modes propagating with different refractive indices gain a phase shift to each other, which results in wave polarization transformation. Since the instantaneous magnitudes of refractive indices in fast and slow axis may vary due to rate of bending or compressing, the resulting phase difference relies on average magnitudes of the refractive indices in both axes.

In order to restore the original polarization state, the wave at the output of the birefringent fiber has to be reflected and coupled back in the fiber together with modes conjugation. To accomplish this, Faraday polarization rotator and flat mirror is exploited, as shown in Fig. 15. The whole device is called ortho-conjugation reflector (OCR) or often Faraday

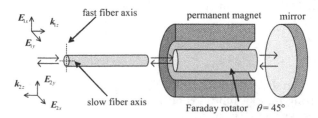

Fig. 15. Principle of Faraday rotation mirror.

rotation mirror (FRM). It consists of Faraday rotator, mounted inside a permanent magnet, flat mirror and collimator (for fiber optic application). The magnetic field magnitude, rotator dimension and its assembly is properly adjusted, that a light wave polarization is rotated with angle 45° while passing the rotator. Mirror allows reflection in perpendicular direction and collimator serves for fiber coupling.

Consider a birefringent fiber with a fast axis and a slow axis, Fig. 15. The modes of light wave, which travels in fiber, gain a phase shift. The mode in slow axis is retarded, while the mode in fast axis travels faster. At the fiber output, the wave is collimated to rotator. It rotates the wave polarization with angle $\theta/2 = 45°$ during a single pass. Then the wave is reflected back. After the second rotator pass, the rotation angle is $\theta = 90°$ due to the rotation non-reciprocity. The wave is coupled back into the fiber. Now, the wave propagates in fiber in backward direction. The wave mode, which traveled previously along the fast axis, travels now along the slow axis. Conversely, the mode, which traveled previously along the slow axis, travels now along the fast axis. The total phase shift of the modes is equalized and the original polarization state is restored. It should be mentioned, although the difference of average refractive indices will no be constant in time (for example by fiber manipulation), the final phase shift will remain zero thanks to reciprocal compensation. The temperature stability of the method is considerable also. However, this is true only when the temperature of FRM is stable, since the Verdet constant of rotator in FRM is temperature dependent.

Though the principle is not applicable for telecommunication purposes, it may be utilized in applications, where the polarization state preservation is desirable. This is often required in erbium-doped fiber amplifiers or tunable fiber lasers. Fiber optic sensors are another field of application. In case of fiber interferometers a polarization state of waves incoming from reference arm and sensing arm has to be preserved in order to interfere. Therefore, FRMs are used in both arms of interferometer. Fiber optic current sensors are another example of usage of FRM (Drexler & Fiala, 2008). Fiber current sensors exploit polarization rotation of the guided wave due to the magnet field actuation. Since the circular birefringence owing to Faraday magneto-optic effect is of non-reciprocal character, it will not be compensated during the backward propagation. Moreover, the polarization rotation will be double, which improves the sensitivity of the sensor. The example of fiber optic polarimetric current sensor utilizing FRM is shown in Fig. 16 (Drexler & Fiala, 2009). Laser beam from laser source L is collimated by means of collimator C and linear polarized by means of polarizer P. Beam passes a non-polarizing beam splitter NBS and it its collimated

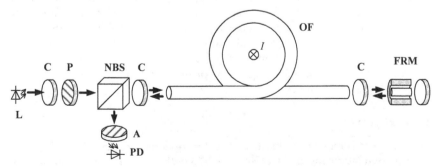

Fig. 16. Fiber optic polarimetric current sensor utilizing Faraday rotation mirror.

into the fiber by collimator C. The beam propagates through the magnetic field sensing fiber OF and exits the fiber. Collimator C collimates the beam into the FRM and collimates backward beam back into the fiber. The beam travels the fiber in opposite direction. After the collimation, part of the beam is deflected by beam splitter NBS. It passes the analyzer A and hits the photodetector PD. The output linear polarization state is perpendicular to the input linear polarization state. When magnetic field acts on the sensing part of the fiber, the polarization plane rotates. The polarization modulation is converted on intensity modulation by means of analyzer A.

In spite of the advantages of the FRM application, several drawbacks limits its usage. One of the drawbacks is temperature dependence of rotator Verdet constant (Santoyo-Mendoza & Barmenkov, 2003). Therefore the FRM unit has to be temperature stabilized. Simultaneously, it has to be shielded from outer magnetic field. Indispensable is the cost of this solution also owing to precise fabrication and adjusting of the FRM. Commercially available are FRMs in compact fiber pigtailed housing for longer wavelengths (1310 nm, 1550 nm). There are also available FRMs for shorter wavelengths (633 nm). However, they are bulky, because of the need of more powerful magnet.

7. Conclusion

Due to their unique properties, single mode fibers have found a huge application potential in various fields of industry and science. They are massively exploited in telecommunication technology, control and sensor systems, industrial laser systems and they represent an unsubstitutable tool for advanced science. The requirements for specific fiber properties differ for various applications. In lot of them, a transmission of light wave with preserving of state of polarization is demanded, which is often a weak point of common fibers. However, this drawback is possible to overcome with a suitable approach, depending on demands of particular application.

In order to evaluate the possibility of polarization state distortion, various influences have to be conceived. It is also very advantageous to be able to quantify them. According to this demand, the intention of the first part of the contribution is to specify the fundamental effects, which lead to fiber birefringence. The basic relations, which allow estimating the birefringence rate are presented also.

Once the fiber birefringence occurs, in lot of cases arises a need to suppress its undesirable consequences. A various approaches, which may be utilized, are the point of the interest of the second part of the contribution. The principles of the most significant methods are described. Their advantages and disadvantages are presented also. The suitability of a selected method is given by the application requirements. They differ in cost, complexity, temperature and mechanical stability and others. Because of the limited extent of this contribution, all of the methods properties and details could not be presented. Nonetheless, the chapter may be a convenient starting point for orientation in this field and details may be found in cited reference sources.

8. Acknowledgement

The contribution has been prepared with support of research project GA102/09/0314 of the Czech Science Foundation and project FEKT-S-10-13 of the Grant Agency of Brno University of Technology.

9. References

Born, M. & Wolf, E. (1999). *Principles of Optics*, Cambridge University Press, ISBN 978-0521642224, Cambridge

Claus, R. O. & Fang, X. (1996). Optimal Design of IRIS-Based Polarimetric Intrinsic Optic Current Sensor. *Journal of Lightwave Technology*, Vol. 14, No. 7, pp. (1664-1673), ISSN 0733-8724

Craig, A. E. & Chang, K. (Ed(s).). (2003). *Handbook of Optical Components and Engineering*, John Wiley & Sons, ISBN 0-471-39055-0, New Jersey

Drexler, P. & Fiala, P. (2008). Utilization of Faraday mirror in fiber optic current sensors. *Radioengineering*, Vol. 17, No. 4, pp. (101-107), ISSN 1210-2512.

Drexler, P. & Fiala, P. (2009). Suppression of polarimetric birefringence effect in optical fiber and its application for pulsed current sensing, *Proceedings of 2009 Waveform Diversity & Design Conference*, ISBN 978-1-4244-2971- 4, USA: Orlando, 2009

Fang, X., Wang, A., May, R. G. & Claus, R. O. (1994). A Reciprocal-Compensated Fiber Optic Current Sensor. *Journal of Lightwave Technology*, Vol. 12, No. 10, pp. (1882-1890), ISSN 0733-8724

Higuera-Lopez, J. M. (2002). *Handbook of Optical Fiber Sensing Technology*, John Wiley & Sons, ISBN 978-0-471-82053-6, New York

Huard, S. (1997). *Polarization of Light*, John Wiley & Sons, ISBN 2-225-85327-X, Paris

Iizuka, K. (2002). *Elements of Photonics, Volume II*, John Wiley & Sons, ISBN 0-471-83938-8, Toronto

Jones, R. C. (1941). New calculus for the treatment of optical systems. *Journal of the Optical Society of America*, Vol. 31, No. 7, ISSN 1862-6254

Kaminow, I. P. and Ramaswamy, V. (1979). Single-polarization optical fibers: slab model. *Applied Physics Letters*, No. 34, pp. (268-270), ISSN 0003-6951

Laming, R. I. & Payne, D. N. (1989). Electric current sensors employing spun highly birefringent optical fibers. *Journal of Lightwave Technology*, Vol. 7, No. 12, pp. (2084-2094), ISSN 0733-8724

Namihira, Y. (1983). Opto-elastic constants in single-mode optical fibers. *Journal of Lightwave Technology*, Vol. 3, No. 5, pp. (1078-1083), ISSN 0733-8724

Nikitov, S., Chamorovskiy, Y., Starostin, N., Ryabko, M., Morshnev, S., Morshnev, V. & Vorob'ev, I. (2009). Microstructured optical fibers for the fiber optics sensors, *Proceedings of International Conference on Materials for Advanced Technologies*, Singapore, June 2009

Noda, J., Okamoto, K. & Sasaki, J. (1986). Polarization-maintaining fibers and their applications. *Journal of Lightwave Technology*, Vol. 4, No. 8. pp. (1071-1089), ISSN 0733-8724

Payne, D. N., Barlow, A. J. & Hansen, J. J. R. (1982). Development of Low- and High-Birefringence Optical Fibers. *IEEE Transaction on Microwave Theory and Techniques*, Vol. 30, No. 4, pp. (323-334), ISSN 0018-9480

Ripka, P. (Ed(s).) (2001). *Magnetic sensors and magnetometers*, Artech House, ISBN 1-58053-057-5, London

Rose, A., Ren, Z. F. & Day, G. W. (1996). Twisting and annealing optical fiber for current sensors. *Journal of Lightwave Technology*, Vol. 14, No. 11, pp. (2492-2498), ISSN 0733-8724

Saleh, B. A. & Teich, M. C. (1991). *Fundamentals of Photonics*, John Wiley & Sons, ISBN 978-0471839651, New York

Santoyo-Mendoza, F. & Barmenkov, Y. O. (2003). Faraday plasma current sensor with compensation for reciprocal birefringence induced mechanical perturbations. *Journal of Applied Research and Technology*, Vol. 1, No. 2, pp. (157–163), ISSN 1665-6423

Senior, J. M., (2009). *Optical Fiber Communication: Principles and Practice*, Pearson Education, ISBN 978-0-13-032681

Stone, J. (1988). Stress-optic effect, birefringence, and reduction of birefringence by annealing in fiber Fabry-Perot interferometers. *Journal of Lightwave Technology*, Vol. 6, No. 7, pp. (1245-1248), ISSN 0733-8724

Tabor, W. J. & Chen, F. S. (1968). Electromagnetic Propagation Through Materials Processing both Faraday Rotation and Birefringence: Experiments with Ytterbium Orthoferrite. *Journal of Applied Physics*, Vol. 10, pp. (2760-2765), ISSN 0021-8979

Ulrich, R., Rashleigh, S. C. & Eickhoff, W. (1980). Bending-induced birefringence in single-mode fibers. *Optical Letters*, Vol. 5, No. 5, pp. (273 – 275), ISSN 1539-4794

Wagner, E., Dändliker, R. & Spenner, K. (Ed(s).). (1992). *Sensors, a comprehensive survey, Volume 6 – Optical Sensors*, VCH, ISBN 3-527-26772-7, Weinheim

Waynant, R. & Ediger, M. (2000). *Electro-Optics Handbook*, McGraw-Hill Professional, ISBN 978-0-07-068716-5, New York

Wilsch, M., Menke, P. & Bosselman, T. (1996). Magneto-optic current transformers for applications in power industry, *Proceedings of 2nd Congress of Optical Sensor Technology OPTO 1996*, Germany: Leipzig, 1996

7

Polarization Losses in Optical Fibers

Hassan Abid Yasser
Thi-Qar University,
Iraq Republic

1. Introduction

Very long span optical communications are mainly limited by the chromatic dispersion (CD) or group velocity dispersion (GVD), fiber nonlinearities, and optical amplifier noise (Agrawal 2005). Different frequencies of a pulse travel with their own velocities, which involves a pulse spreading. In a fiber-optic communication system, information is transmitted within a fiber by using a coded sequence of optical pulses whose width is determined by the bit rate of the system. The CD induced broadening of pulses is undesirable phenomenon since it interferes with the detection process leading to errors in the received bit pattern (Kogelnik & Jopson 2002; Mechels et al. 1997). Clearly GVD will limit the bit rate and the transmission distance of a fiber-optic communication system. GVD is basically constant over time, and compensation can be set once and forgotten (Karlsson 1994).

When the signal channel bit rates reached beyond 10 Gb/s, polarization mode dispersion (PMD) becomes interesting to a larger technical community. PMD is now regarded as a major limitation in optical transmission systems in general, and an ultimate limitation for ultra-high speed signal channel systems based on standard single mode fibers (Mahgerftech & Menyuk 1999). PMD arises in optical fibers when the cylindrical symmetry is broken due to noncircular symmetric stress. The loss of such symmetry destroys the degeneracy of the two eigen-polarization modes in fiber, which will cause different GVD parameters for these modes. In standard single mode fibers, PMD is random, i.e. it varies from fiber to fiber. Moreover, at the same fiber PMD will vary randomly with respect to wavelength and ambient temperature (Lin & Agrawal 2003b; Sunnerud et al. 2002). The differential group delay (DGD) between two orthogonal states of polarization called the principal states of polarization (PSP's) causes the PMD (Tan et al. 2002; Wang et al. 2001). As a pulse propagates through a light-wave transmission system with a PMD, the pulse is spilt into a fast and slow one, and therefore becomes broadened. This kind of PMD is commonly known as first-order PMD. Under first-order PMD, a pulse at the input of a fiber can be decomposed into two pulses with orthogonal states of polarization (SOP). Both pulses will arrive at the output of the fiber undistorted and polarized along different SOP's, the output SOP's being orthogonal (Chertkov et al. 2004; Foshchini & Poole 1991). Both the PSP's and the DGD are assumed to be frequency independent when only first-order PMD is being considered (Lin & Agrawal 2003c; Gordon & Kogelnik 2000).

Second-order PMD effects account for the frequency dependence of the DGD and the PSP's. The frequency dependence of the DGD introduces an effective chromatic dispersion of opposite sign on the signals polarized along the output PSP's (Elbers et al. 1997; Ibragimv &

Shtenge 2002). Fiber PMD causes a variety of impairments in optical fiber transmission systems. First of all there is the inter-symbol interference (ISI) impairment of a single digital transmission channel. The ISI impairment is caused by the DGD between the two pulses propagating in the fiber when the input polarization of the signal does not match one of the PSP's of the fiber PMD impairments due to inter-channel effects that occur in polarization-multiplexed transmission systems (Agrawal 2005; Yang et al. 2001).

There are two polarization effects that lead to impairments in the long-haul optical fiber transmission systems: PMD and polarization dependent loss (PDL) (Chen et al. 2003; Chen et al. 2007). The WDM systems whose channels are spread over a large bandwidth rapidly change their state of polarizations (SOP's) due to PMD so that the overall DOP of the system is nearly zero (Agrawal 2005; Kogelnik & Jopson 2002). At the same time different channels expiries different amounts of PDL, and since the amplifiers maintain the total signal power nearly constant, individual channels undergo a kind of random walk so that it is possible for some channels to fade (Shtaif & Rosenberg 2005; Menyuk et al. 1997). Calculating the impairments due to the combination of PMD and PDL in WDM systems is a formidable theoretical challenge (Phua & Ippen 2005). Physically, light pulses polarized along these PSP's propagate without polarization-induced distortion. When there is no PDL, the two PSP's are orthogonal and correspond to the fastest and slowest pulses, which can propagate in the fiber (Yasser 2010; Yaman et al. 2006). They thus constitute a convenient basis for polarization modes. When the system includes PDL, the Jones formalism is still applicable, but several of the above facts are not valid anymore. The notion of PSP's is still correct, but the two PSP's are not orthogonal nor do they represent the fastest and slowest pulses (Yoon & Lee 2004).

In this chapter, the analysis of Jones and Stokes vectors and the relation between them were discussed in section 2. The statistics of PMD are presented in section 3. The pulse broadening in presence of PMD and CD were illustrated in section 4. In section 5, the principal comparison between PMD and birefringence vector will be obtained. The combined effects of PMD and PDL are presented in section 6. Finally, section 7 will summarize the effects of nonlinearity on the effective birefringence vector.

2. Polarization dynamics

The representation of polarization in Jones and Stokes spaces and the connection between the two spaces will be presented in this section. Throughout this chapter, it is assumed that the usual loss term of the fiber has been factored out so that one can deal with unitary transmission matrices. Light in optical fibers can be treated as transverse electromagnetic waves. Considering the two perpendicular and linearly polarized light waves propagating through the same region of space in the z-direction, the two fields can be represented in complex notation as (Azzam & Bashara 1989)

$$\vec{E}_x(z,t) = \hat{x}\, E_{x0}\, e^{i(kz-wt+\varphi_x)} \tag{1a}$$

$$\vec{E}_y(z,t) = \hat{y}\, E_{y0}\, e^{i(kz-wt+\varphi_y)} \tag{1b}$$

where φ_x and φ_x are the phases of the two field components, and k is the propagation constant. The resultant optical field is the vector sum of these two perpendicular waves, i.e.

$$\vec{E}(z,t) = \vec{E}_x(z,t) + \vec{E}_y(z,t) \tag{2}$$

The polarization state can be represented in terms of Jones vectors as

$$\hat{A} = \begin{bmatrix} a_x e^{i\varphi_x} \\ a_y e^{i\varphi_y} \end{bmatrix} \tag{3}$$

where $a_x = E_{x0} / \sqrt{E_{x0}^2 + E_{y0}^2}$, $a_y = E_{y0} / \sqrt{E_{x0}^2 + E_{y0}^2}$, and $\sqrt{a_x^2 + a_y^2} = 1$. Here E_{x0} and E_{y0} are the initial amplitude components of the light. The familiar form of Jones vector is denoted as ket vector as (Gordon & Kogelnik 2000)

$$|s> = \begin{bmatrix} s_x \\ s_y \end{bmatrix} = \begin{bmatrix} a_x e^{i\varphi_x} \\ a_y e^{i\varphi_y} \end{bmatrix} \tag{4}$$

whereas the bra $<s|$ indicates the corresponding complex conjugate row vector, i.e. $<s| = [s_x^* \ s_y^*]$, where * indicates complex conjugation. The bra-ket notation is used to distinguish Jones vectors from another type of vectors that will be used in this chapter which is called the Stokes vectors. Partial correlation yields partial polarization and total correlation gives total polarization (Karlsson 1994; Sunnerud et al. 2002). When the light is coherent, Jones vectors are all of unit magnitude, i.e. $<s|s> = s_x s_x^* + s_y s_y^* = 1$. Given the Jones vector, the values of the azimuth angle, ψ, and the ellipticity angle, η, can be found by solving the equations (Rogers 2008)

$$\tan 2\psi = \frac{2\text{Re}(s_y / s_x)}{1 - |s_y / s_x|^2} \tag{5a}$$

$$\sin 2\eta = \frac{2\text{Im}(s_y / s_x)}{1 + |s_y / s_x|^2} \tag{5b}$$

where Re and Im denote the real and imaginary parts, respectively. Fig.(1 a) illustrates Jones representation of polarization vector.

The Poincare sphere is a graphical tool in real three dimensional space that allows convenient description of polarized signals and polarization transformations caused by propagation through devices. Any SOP can be represented uniquely by a point on or within a unit sphere centered on a rectangular coordinates system. The coordinates of a point are the three normalized Stokes parameters describing the state of polarization (Azzam & Bashara 1989; Rogers 2008). Partially polarized light can be considered as a combination of purely polarized light and un-polarized light. Orthogonal polarizations are located diametrically opposite to the sphere. As shown in Fig.(1 b), linear polarizations are located on the equator. Circular states are located at the poles, with intermediate elliptical states continuously distributed between the equator and the poles (Karlsson 1994; Kogelnik & Jopson 2002). There are two angles (or degrees of freedom, i.e. ψ and η) describing an arbitrary Jones vector. These angles can be interpreted as coordinates in a spherical coordinates system, and each polarization state can then correspond to a point, represented by a Stokes vector, $\hat{s} = (s_1, s_2, s_3)^t$ on the Poincare sphere, where t represents the transpose. The three Cartesian components can be defined as (Gordon & Kogelnik 2000)

$$s_1 = |E_x|^2 - |E_y|^2 = \cos 2\psi \cos 2\eta$$
$$s_2 = E_x E_y^* + E_y E_x^* = \sin 2\psi \cos 2\eta \qquad (6)$$
$$s_3 = i(E_y E_x^* - E_x E_y^*) = \sin 2\eta$$

Therefore, the angle 2ψ is the angle from the direction of s_1 to the projection of $\hat{s}$ on the $s_1 - s_2$ plane, and 2η is the angle from $s_1 - s_2$ plane to the vector $\hat{s}$, see Fig. (1 b). Given Stokes vector, the values of ψ and η are obtained by solving the equations $s_2 / s_1 = \tan 2\psi$, and $s_3 = \sin 2\eta$.

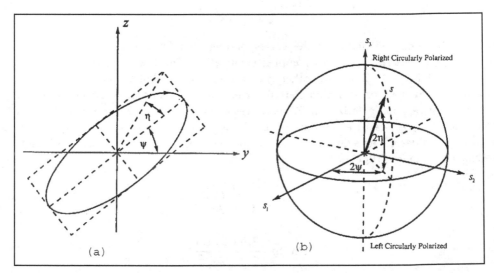

Fig. 1. Illustration of: a) Jones representation, b) Stokes representation.

Any Stokes vector $\hat{s}$ is related to another one $|s>$ in Jones space as $\hat{s} = <s|\vec{\sigma}|s>$, where $\vec{\sigma} = (\sigma_1, \sigma_2, \sigma_3)$ is the Pauli spin vector whose components are defined as (Levent et al. 2003)

$$\sigma_1 = \begin{bmatrix} 1 & 0 \\ 0 & -1 \end{bmatrix}, \quad \sigma_2 = \begin{bmatrix} 0 & 1 \\ 1 & 0 \end{bmatrix}, \quad \sigma_3 = \begin{bmatrix} 0 & -i \\ i & 0 \end{bmatrix} \qquad (7)$$

It is important to note that if the angle between $\hat{p}$ and $\hat{s}$ in Stokes vector is θ, then the angle between $|p>$ and $|s>$ in Jones space is $\theta / 2$. That is; if two vectors are perpendicular in Jones space then the corresponding two vectors in Stokes space are antiparallel. Each of these two spaces gives certain illustrations according to the case of study. For totally polarization, the value of polarization vector is unity, elsewhere, the value differs from unity. In general, the three components of $\hat{s}$ are not zero for elliptical polarization. The third component of $\hat{s}$ equals zero for linear polarization, whereas the first two components of $\hat{s}$ are zero for circular polarization. There is a unitary matrix, T, in Jones space which relates output to input via $|s>=T|t>$, where $|s>$ and $|t>$ are the output and input Jones vectors, respectively. On the other hand, a transformation matrix (Muller), R, in Stokes space relates output to input via $\hat{s} = R\,\hat{t}$, where $\hat{s}$ and $\hat{t}$ are the output and input Stokes vectors, respectively. The transmission matrices are related as $R\vec{\sigma} = T^{\dagger}\vec{\sigma}T$, where $\dagger$ denotes the transpose of the complex conjugate (Agrawal 2007; Chen et al. 2007).

3. Statistical managements

The effects of PMD are usually treated by means of the three-dimensional PMD vector that is defined as $\vec{\tau} = \tau_{pmd}\hat{p}$, where $\hat{p}$ is a unit vector pointing in the direction of slow PSP and τ_{pmd} is the DGD between the fast and slow components which is defined as (Mahgerftech & Menyuk 1999)

$$\tau_{pmd} = |\vec{\tau}| = \sqrt{\tau_1^2 + \tau_2^2 + \tau_3^2} \tag{8}$$

The PMD vector $\vec{\tau}$ in Stokes space gives the relation between the output SOP, $\hat{s}$, and the frequency derivative of the output SOP: $d\,\hat{s}(w)\,/\,dw = \vec{\tau}(w)\times\hat{s}(w)$. The PSP's are defined as the states that $\vec{\tau}(w)\times\hat{s}(w)=0$, so that no changes in output polarization can be observed close to these states at first order in w. To the first order, the impulse response of an optical fiber with PMD is defined as (Karlsson 1994)

$$h_{pmd}(T) = \gamma_{+}\delta(T - \tau_{pmd}\,/\,2)\,|\,p_{+}> +\gamma_{-}\delta(T + \tau_{pmd}\,/\,2)\,\,|\,p_{-}> \tag{9}$$

where $\gamma_{\pm}$ are the splitting ratios and $|p_{\pm}>$ are the PSP's vectors. The factors $\gamma_{\pm}$ and τ_{pmd} vary depending on the particular fiber and its associated stresses, where the splitting ratios can range from zero to one. Note that, the function $h_{pmd}(T)$ is normalized in the range ($-\infty$ to ∞).

3.1 Splitting ratios

Consider that the PSP's occur with a uniform distribution over the Poincare sphere, and that $\hat{s}$ is aligned with the north pole of the sphere as shown in Fig.(2). The probability density of PSP's which is found in the range $d\theta$ about the angle θ relative to $\hat{s}$ is proportional to the differential area $2\pi\sin\theta\,d\theta$ sketched in the figure. As there is north/south symmetry in the differential area, the ranges (0 to $\pi/2$) and ($\pi/2$ to π) of θ are combined to obtain the combined probability density $p_{\theta}(\theta) = \sin\theta$. For the effective range (0 to $\pi/2$) describing the occurrence of PSP's with angle θ (and $\pi-\theta$) relative to $\pm\hat{s}$, the distribution $p_{\theta}(\theta)$ is properly normalized through the range (0 to $\pi/2$). The analyses of splitting ratios have led to a number of important fundamental advances as well as the technical point of view (Rogers 2008; Kogelnik & Jopson 2002). The splitting ratios $\gamma_{\pm}$ can be determined from the polarization vectors. In other words $\gamma_{\pm}$ represent the projection of $|p_{+}>$ and $|p_{-}>$ onto $|s>$. Formally, $\gamma_{\pm}^2 = |<s|p_{\pm}>|^2$, where $|s>$ and $|p_{\pm}>$ are the input SOP and the two PSP's vectors. If the PSP's are defined as $|p_{\pm}>=[p_{\pm x}\ p_{\pm y}]^{t}$, then

$$|p_{\pm}><p_{\pm}| = \begin{bmatrix} p_{\pm x} \\ p_{\pm y} \end{bmatrix}\begin{bmatrix} p_{\pm x}^{*} & p_{\pm y}^{*} \end{bmatrix} = \begin{bmatrix} |p_{\pm x}|^2 & p_{\pm x}p_{\pm y}^{*} \\ p_{\pm y}p_{\pm x}^{*} & |p_{\pm y}|^2 \end{bmatrix} \tag{10}$$

where $< p_\pm |$ are the transpose conjugation of $| p_\pm >$. Now, it is straightforward to show that

$$\pm \hat{p} = < p_\pm | \vec{\sigma} | p_\pm > = \begin{bmatrix} \pm p_1 \\ \pm p_2 \\ \pm p_3 \end{bmatrix} = \begin{bmatrix} < p_\pm | \sigma_1 | p_\pm > \\ < p_\pm | \sigma_2 | p_\pm > \\ < p_\pm | \sigma_3 | p_\pm > \end{bmatrix} = \begin{bmatrix} | p_{\pm x} |^2 - | p_{\pm y} |^2 \\ p_{\pm x} p_{\pm y}^* + p_{\pm y} p_{\pm x}^* \\ i(p_{\pm y} p_{\pm x}^* - p_{\pm x} p_{\pm y}^*) \end{bmatrix} \tag{11}$$

Comparing Eqs.(10) and (11), $| p_\pm >< p_\pm | = (I_2 \pm \hat{p} \cdot \vec{\sigma}) / 2$ can be extracted. In turn, the splitting ratios can be calculated by using Eq.(11) and the fact that $< a | \hat{p} \cdot \vec{\sigma} | a >= \hat{p} \cdot \hat{a}$ as follows

$$\gamma_+^2 = < s | p_+ >< p_+ | s >= < s | (I_2 + \hat{p} \cdot \vec{\sigma}) | s > / 2 = (1 + \hat{p} \cdot \hat{s}) / 2 = \cos^2(\theta / 2) \tag{12a}$$

$$\gamma_-^2 = < s | p_- >< p_- | s >= < s | (I_2 - \hat{p} \cdot \vec{\sigma}) | s > / 2 = (1 - \hat{p} \cdot \hat{s}) / 2 = \sin^2(\theta / 2) \tag{12b}$$

Until now, the relationship between the splitting ratios and elevation angle was calculated, where the ratios $\gamma_\pm$ are identical only for $\theta = \pi / 2$.

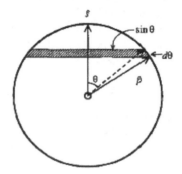

Fig. 2. Sketch of differential area on Poincare sphere as a function of elevation angle θ.

3.2 Statistics of DGD

Throughout this subsection, the PMD statistics have been carefully analyzed since it causes a variation in the pulse properties. A proper measure of pulse width for pulses of arbitrary shapes is the root-mean square (rms) width of the pulse defined as $\tau_{rms} = \sqrt{< T^2 > - < T >^2}$. The PMD induced pulse broadening is characterized by the rms value of τ_{pmd}. The τ_{rms} is obtained after averaging over random birefringence changes. The second moment of τ_{pmd} is given by (Fushchini & Poole 1991)

$$< \tau_{pmd}^2 >= \tau_{rms}^2 = 2(\Delta \beta_1)^2 \, \ell_c^2 \, [L / \ell_c + e^{-L/\ell_c} - 1] \tag{13}$$

where ℓ_c is the correlation length that is defined as the length over which two polarization components remain correlated, $\Delta \beta_1 = v_{gx}^{-1} - v_{gy}^{-1}$ is related to the difference in group velocities along the two PSP's. For distances $L >> 1 \, km$, a reasonable estimate of pulse broadening was obtained by taking the limit $L >> \ell_c$ in Eq.(13). The result is given by $\tau_{rms} \approx \Delta \beta_1 \sqrt{2 L \ell_c} = D_p \sqrt{L}$, where D_p is known as the PMD parameter that takes the values $(0.01-10) ps / \sqrt{km}$. The variable τ_{pmd} has been determined to obey a Maxwellian distribution of the form (Agrawal 2005)

$$p(\tau_{pmd}) = \sqrt{\frac{54}{\pi}} \frac{\tau_{pmd}^2}{\tau_{rms}^3} e^{-3\tau_{pmd}^2/2\tau_{rms}^2} \tag{14}$$

The mean of τ_{pmd} is done simply as $\bar{\tau}_{pmd} = \sqrt{8/3\pi}\, \tau_{rms}$. Using this result, the Maxwellian distribution will take the form

$$p(\tau_{pmd}) = \frac{32}{\pi^2} \frac{\tau_{pmd}^2}{\bar{\tau}_{pmd}^3} e^{-4\tau_{pmd}^2/\pi\bar{\tau}_{pmd}^2} \tag{15}$$

A cursory inspection of Eq.(15) reveals that the $p(\tau_{pmd})$ can be found if $\bar{\tau}_{pmd}$ is known. Here, a relationship for τ_{pmd} that will maximize $p(\tau_{pmd})$ can be found. The distribution $p(\tau_{pmd})$ has a maximum value at $\tau_{pmd} = \tau_{pmd}^{max} = \sqrt{\pi}\bar{\tau}_{pmd}/2$. This conclusion provides a method for calculating the maximum likelihood value of τ_{pmd} if $\bar{\tau}_{pmd}$ is known.

3.3 Statistics of impulse response

The rms width of the impulse response, τ_{eff}, can be readily calculated by substituting Eq.(9) into $\tau_{rms} = \sqrt{<T^2> - <T>^2}$ to yield

$$\tau_{eff} = \sqrt{\int_{-\infty}^{\infty} T^2 h_{pmd}(T)dT - \left[\int_{-\infty}^{\infty} Th_{pmd}(T)dT\right]^2} = \sin\theta \, \tau_{pmd}/2 \tag{16}$$

Using the result $p_\theta(\theta) = \sin\theta$ and Eq.(16), the density distribution for θ can be transformed to the density for τ_{eff} as follows

$$p_{\tau_{eff}}(\tau_{eff}) = p_\theta(\theta(\tau_{eff})) \frac{d\theta}{d\tau_{eff}} = \frac{4\tau_{eff}}{\tau_{pmd}\sqrt{\tau_{pmd}^2 - 4\tau_{eff}^2}} \tag{17}$$

It is important to note that the probability density is a function of τ_{eff} and τ_{pmd}. As a consequence of this dependence, Eq.(17) can not be integrated to determine τ_{eff} due to the presence of the other variable τ_{pmd}. So, the next step is to seek about $p_{\tau_{eff}}(\tau_{eff})$ in order to determine the statistical properties of output pulses. The joint probability distribution $p(\tau_{eff}, \tau_{pmd})$ can be illustrated using Eqs.(14) and (17) as follows

$$p(\tau_{eff}, \tau_{pmd}) = \frac{64\tau_{eff}}{\pi\bar{\tau}_{pmd}^3} \frac{\tau_{pmd}}{\sqrt{\tau_{pmd}^2 - 4\tau_{eff}^2}} e^{-4\tau_{pmd}^2/\pi\bar{\tau}_{pmd}^2} \tag{18}$$

Recalling Eq.(16), it may be written as $\tau_{pmd} = 2\tau_{eff}/\sin\theta$. Since $0 \le \sin\theta \le 1$, such that $2\tau_{eff} \le \tau_{pmd} < \infty$. The probability distribution $p(\tau_{eff})$ can be found by integrating Eq.(18) about τ_{pmd} through the range $2\tau_{eff} \le \tau_{pmd} < \infty$ to obtain

$$p(\tau_{eff}) = \frac{32\tau_{eff}}{\pi\bar{\tau}_{pmd}^2} e^{-16\tau_{eff}^2/\pi\bar{\tau}_{pmd}^2} \tag{19}$$

At a basic level, Eq.(17) is the same as Eq.(19) but the latter is a function of τ_{eff} only, which can be integrated to obtain τ_{eff}. However, both equations are normalized properly. The mean value of τ_{eff} is determined as $\theta = \pi / 2$. So, Eq.(19) may be written as

$$p(\tau_{eff}) = \frac{\pi}{2} \frac{\tau_{eff}}{\tau_{eff}^2} \, e^{-\pi \, \bar{\tau}_{eff}^2 / 4 \, \bar{\tau}_{eff}^2} \tag{20}$$

The distribution $p(\tau_{eff})$ has a maximum value at $\tau_{eff} = \tau_{eff}^{max} = \sqrt{\pi / 32} \bar{\tau}_{pmd}$. This is equivalent to find the maximum likelihood value of τ_{eff} if $\bar{\tau}_{pmd}$ is known.

3.4 Pulse characteristics

Using the PSP's as an orthogonal basis set, any input or output polarization can be expressed as the vector sum of two components, each aligned with a PSP. Within the realm of the first-order PMD, the output electric field from a fiber with PMD has the form (Rogers 2008)

$$| A_{out}(T) > = \gamma_+ \, A_{in}(T - \tau_{pmd} / 2) \ | p_+ > + \gamma_- \, A_{in}(T + \tau_{pmd} / 2) \ | p_- > \tag{21}$$

where $A_{in}(T)$ is the input electric field. To determine the output power $P_{out}(T) = < A_{out}(T) | A_{out}(T) >$, it is important to point out the orthogonality properties of Jones vectors, that is; $< p_\pm | p_\mp > = 0$ and $< p_\pm | p_\pm > = 1$. Note that, we perform the derivation using a normalized Gaussian pulse that takes the form $A_{in}(T) = D\exp(-T^2 / 2T_0^2)$, where $D = \sqrt{E_{in} / T_0 \sqrt{\pi}}$, T_0 is the initial pulse width, and E_{in} is the input pulse energy. For normalized power, we make $D^2 = 1$. Therefore, according to Eq.(21), the shifted pulses will reshape as

$$A_{in}(T \pm \tau_{pmd} / 2) = D\exp\left[-\frac{(T \pm \tau_{pmd} / 2)^2}{2T_0^2} \right] \tag{22}$$

Substituting Eqs.(12) and (22) into (21), using the output power definition, using the orthogonality properties of Jones vectors, and simplified the result, we obtain the following expression

$$P_{out}(T) = \left[\cos^2(\theta \ / 2) \, e^{-T\tau_{pmd}/T_0^2} + \sin^2(\theta / 2) \, e^{T\tau_{pmd}/T_0^2} \right] e^{-(4T^2 + \tau_{pmd}^2)/4T_0^2} \tag{23}$$

The width of the output pulse T_1 can be determined as follows

$$T_1 = \sqrt{\int_{-\infty}^{\infty} T^2 P_{out}(T)dT - \left[\int_{-\infty}^{\infty} TP_{out}(T)dT \right]^2} = \sqrt{T_0^2 + (\tau_{pmd} / 2)^2 \sin^2 \theta} \tag{24}$$

The time jittering of the pulse can be found by determining the maximum value of $P_{out}(T)$. This maximum value will happen at $T = T_{peak} = \tau_{pmd} \cos(\theta) / 2$. The peak power, as a function of DGD and an angle θ, at the pulse center can be determined by substituting the latter result into Eq.(23) to get

$$P_{peak}(\tau_{pmd}, \theta) = \cos^2(\theta/2) \, e^{-\sin^2(\theta/2)\tau_{pmd}^2/T_0^2} + \sin^2(\theta/2) \, e^{-\cos^2(\theta/2)\tau_{pmd}^2/T_0^2} \tag{25}$$

At this point, we drive formulas for the output power form, final width, time jittering (shifting), and peak power as functions of the random physical variables θ and τ_{pmd}.

Fig.(3) illustrates the simulation with the parameters: $L = 50\ km$, $D_p = 0.5\ ps / \sqrt{km}$, and $T_0 = 5\ ps$. The solid line represents the original pulse while the discrete lines represent the resulted pulses with different values of τ_{pmd} ranging from 0 to $8ps$, where the closest to $T = 0$ is the pulse that has least value of $\bar{\tau}$. At the angle $\theta = 0$, one note that the pulse is faced only by a displacement to the right at $T_{peak} = \tau_{pmd} / 2$. Increasing θ, the pulse width and distortion will be increased, while the power and shifting will be decreased. These variations are the greatest at $\theta = \pi / 2$. After $\theta = \pi / 2$, the effects are reversed. At $\theta = \pi$, again the pulse is faced only by a displacement but to the left at $T_{peak} = -\tau_{pmd} / 2$. It is clear that the penalty could be greater if $\theta = \pi / 2$ and will be zero at $\theta = 0\ or\ \pi$.

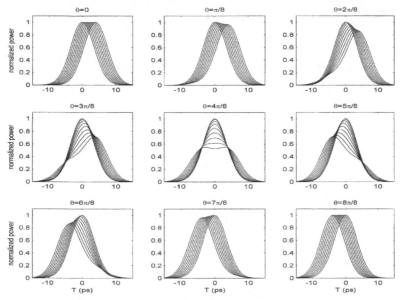

Fig. 3. Pulse shape with different values of τ_{pmd} and θ for different values of τ_{pmd}; the lower value of τ_{pmd} is the closest to the pulse center.

4. Polarization mode dispersion and chromatic dispersion

The pulses that propagate through single mode fiber (SMF) are affected by two types of dispersion which are CD and PMD. Notice that the effects of the two types of dispersion happen at the same time, so to give a distinct sense of the two types of dispersion we decided to obtain the effects in the frequency domain. The initial pulse, $\tilde{A}(0,w) = \Im\{A(0,T)\}$, first faces the affect of CD (the transfer function $H_1(w)$) to obtain $H_1(w)\tilde{A}(0,w)$. The CD does not depend on SOP therefore the input SOP (the Jones vector $|a>$) will not change. Next, the pulse divides into two orthogonal components towards PSP's ($|a^+>$ and $|a^->$) under the effects of PMD. The component in the direction $|a^+>$ will face the effects of the function $H_{2s}(w)$ to obtain the pulse $H_{2s}(w)H_1(w)\tilde{A}(0,w)$ and at the same time the SOP will change from $|a^+>$ to $|b^+>$. On the other hand, the pulse in the direction $|a^->$ faces the effects of the function $H_{2f}(w)$ to yield $H_{2f}(w)H_1(w)\tilde{A}(0,w)$ and also the SOP will change from $|a^->$ to $|b^->$. The input or output PSP's does remain orthogonal when the PDL is absent. Finally, the vector sum of the two components will produce the final pulse $H_2(w)H_1(w)\tilde{A}(0,w)$. The transfer function of the CD of lossless fiber in frequency domain is $H_1(w)=\exp(i\,w^2\,\beta_2\,L/2)$, where $\beta_2=-\lambda^2 d(\lambda)/2\pi\,c$, $d(\lambda)$ is fiber chromatic dispersion parameter, L is the fiber length, and λ is light wavelength. Now, assume that there is negligible PDL, so that we can use the principal states model (Lin & Agrawal 2003b; Ibragimv & Shtenge 2002; Foschini & Poole 1991) to characterize first-order PMD. Under this model, there exist a pair of orthogonal input PSP's, $|a^+>$ and $|a^->$, and a pair of orthogonal output PSP's, $|b^+>$ and $|b^->$, where all of PSP's are expressed as Jones vectors. If an arbitrary polarized field $\vec{A}_a(t)=A_a(t)\,|a>$ is input to the fiber, this input field can be projected onto the two PSP's as

$$\vec{A}_a(T)=\gamma_+\,A_a(T)\,|a^+>+\gamma_-\,A_a(T)\,|a^->\qquad(26)$$

In terms of first-order PMD, the output field of the fiber takes the form

$$\vec{A}_b(T)=\gamma_+A_b(T-\tau_{pmd}/2)|b^+>+\gamma_-\,A_b(T+\tau_{pmd}/2)\,|b^->\qquad(27)$$

According to Eq.(9), the fiber transfer functions for first-order in the time and frequency domains are given by

$$h_2(t)=\gamma_+\,\delta(T-\tau_{pmd}/2)\,|b^+>+\gamma_-\,\delta(T+\tau_{pmd}/2)|b^->\qquad(28a)$$

$$H_2(w)=\gamma_+e^{iw\tau_{pmd}/2}\,|b^+>+\gamma_-\,e^{-iw\tau_{pmd}/2}\,|b^->\qquad(28b)$$

The root mean square width of this impulse response which can be calculated as

$$<T>=\int_{-\infty}^{\infty}T\,h_2(T)dT=(\gamma_+\,\tau_{pmd}|b^+>-\gamma_-\,\tau_{pmd}\,|b^->)/2$$

$$<T^2>=\int_{-\infty}^{\infty}T^2h_2(T)dT=(\gamma_+\,\tau_{pmd}^2\,|b^+>+\gamma_-\tau_{pmd}^2\,|b^->)/4\qquad(29)$$

$$\tau_{rms}^{\pm}=[\sqrt{<T^2>-<T>^2}]_{\pm}=0$$

where the signs $(+,-)$ mean that the impulse response in directions of $|b^+>$ or $|b^->$, respectively. That is; the width of an impulse response in the direction of PSP's will be zero, while the width in the direction of $|b>$ will be $\tau_{rms}=\sin\theta\,\tau_{pmd}/2$. This represents the extra width that results due to the effects of PMD on the propagated signal. It is clear that, if the input SOP is in direction of PSP's, then the pulse will not suffer any broadening.

The Fourier transformation of the initial pulse takes the form $\tilde{A}(0,w)=D\sqrt{2\pi}T_o\exp(-w^2T_o^2/2)$. The total effects on the pulse shape can be obtained by using the convolution of the transfer functions of the combined PMD and CD with the input Gaussian signal in the time domain, or equivalently by using the inverse Fourier transform as follows

$$A(z,T)=\mathfrak{I}^{-1}\{\tilde{A}(0,w)\cdot H_1(w)\cdot H_2(w)\}=\cos(\theta/2)A_+(z,T)\,|b^+>+\sin(\theta/2)A_-(z,T)\,|b^->\qquad(30)$$

where

$$A_\pm(z,T) = D\frac{T_o}{\chi}\,\exp(-\frac{T_\pm^2}{2\,T_1^2})\;\;\exp(\,i\,\phi_\pm(z,T)\,)$$

$$\chi = \sqrt[4]{T_o^4 + (\beta_2\,z)^2}$$

$$T_\pm = T \pm \tau_{pmd}\,/\,2$$

$$T_1 = \sqrt{T_o^2 + (\beta_2 z\,/\,T_o^2)^2}$$

$$\phi_\pm(z,T) = -\frac{\beta_2 z}{2T_1^2}\frac{T_\pm^2}{T_o^2} + \frac{1}{2}\tan^{-1}(\beta_2 z\,/\,T_o^2)$$

The parameter T_1 represents the pulse width including CD effects where it is the same for the two orthogonal components. The width of each component will not increase under the effects of PMD, but the pulse which results from the vector sum of the two orthogonal components will face a broadening that can be determined by τ_{rms}. The parameters $\phi_\pm(z,T)$ represent the nonlinear phases that generate through the propagation in optical fiber. The nonlinear phase as a function of time differs from one component to another by the amount τ_{pmd}, but in the frequency domain they remain the same and add the same value of noise to both components. The frequency chirp can be written as

$$\delta\,w_\pm(T) = -\frac{\partial\phi_\pm(z,T)}{\partial T} = \frac{\beta_2 z}{T_1^2}\frac{T_\pm}{T_o^2} = \frac{\beta_2 z}{2T_1^2}\frac{T \pm \Delta\tau\,/\,2}{T_o^2} \tag{31}$$

This means that the new frequencies generated are similar for the two components and the difference lies in $T \pm \tau_{pmd}\,/\,2$ only, which means that one of the components advances the other by time τ_{pmd}. Eq.(30) explains that the pulse amplitude will decrease by increasing the propagation distance, which will be converted to the same equations as in reference (Agrawal 2007) by ignoring the effects of PMD. The Jones vectors $|b^+>$ and $|b^->$ are orthogonal, i.e. $<b^+\,|\,b^->=0$. That is enough to assume a random form to one of them to find the other. For example, if $|b^+>=[x\;\;iy]^t$ then $|b^->=[iy\;\;x]^t$ keeping in mind that all the polarization vectors have unit values.

Now, the reconstructed width after including the effects of CD is T_1. Next, the input pulse has a width T_1 which will be increased by the amount τ_{rms} due to the PMD. Such that, the final width will be

$$T_f = \sqrt{T_1^2 + \tau_{pmd}^2\sin^2\theta\,/\,4} \tag{32}$$

Fig.(4 a) illustrates the shape of pulse for various values of β_2, assuming $\tau_{pmd}=2\;ps$, $\theta = \pi\,/\,2$, $T_o = 10\;ps$, and $L = 60\;km$. Since τ_{pmd} is constant for all cases, this implies that the time separation between the orthogonal components remains the same. The width of both components increases (under the effects of CD) by increasing β_2. Consequently, the width of the final pulse increases by increasing β_2, but the amplitude is decreased. The existence of CD causes a broadening factor (BR) of value $T_1\,/\,T_o$, and the existence of PMD adds a BR of value $\tau_{rms}\,/\,T_1$. That is; the width of pulse will increase due to the existence of

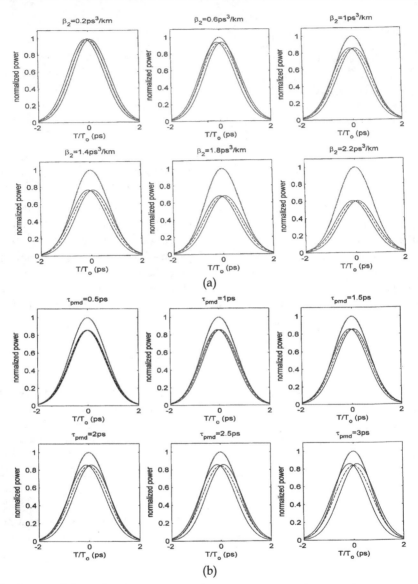

Fig. 4. Evolution of the pulse shape at $\theta = \pi / 2$, $T_0 = 10$ ps, and $L = 60$ km:
a) for various values of β_2 and $\tau_{pmd} = 2$ ps, b) for various values of τ_{pmd} and
$\beta_2 = 1$ ps^3 / km. The dotted, continuous, and discrete lines refer to the initial pulse,
two orthogonal components, and final pulse, respectively.

the two types of dispersion. In other words, the time separation between the two orthogonal
components will be fixed, both amplitude and width of the pulse will change under the
effects of CD. Fig.(4 b) illustrates the shape of pulse for various values of τ_{pmd},
assuming $\beta_2 = 1$ ps^3 / km, $\theta = \pi / 2$, $T_0 = 10$ ps, and $L = 60$ km. Since β_2 and L are

constants this implies that T_1 is constant also. That is; the width of both orthogonal components are similar for all τ_{pmd} values, but the difference appears as a time increase separation between the two components. This leads to adding a BR of value τ_{rms} / T_1 to the reconstructed pulse.

5. Polarization mode dispersion and birefringence

In the optical fibers, the birefringence vector $\vec{\beta}$ may be defined in two forms as (Schuh et al. 1995)

$$\vec{\beta}_L = \begin{bmatrix} \Delta\beta\cos 2\alpha \\ \Delta\beta\sin 2\alpha \\ 0 \end{bmatrix} \quad or \quad \vec{\beta}_{NL} = \begin{bmatrix} \Delta\beta\cos 2\alpha \\ \Delta\beta\sin 2\alpha \\ \zeta T \end{bmatrix} \tag{33}$$

where α is the angle of birefringence in Jones space, $\Delta\beta$ is the magnitude of linear birefringence, i.e. $\Delta\beta = |\vec{\beta}_L|$, ζ is the photo-elastic coefficient of glass, and T is the twist rate in (rad/m). The angle α is not constant along the fiber; also, $\Delta\beta$ and T . This means that each segment of fiber has a birefringence vector differs from another position randomly, depending on the values of α , $\Delta\beta$, and T . If $\Delta\beta = |\vec{\beta}_L|$, then $\vec{\beta}_L - \Delta\beta\,\hat{r}$, where $\hat{r}$ represents a unit vector in Stokes space. The vector $\hat{r}$ represents a rotation axis of the polarization vector, which differs from one section to another randomly.

Consequently, the PMD vector can be defined as a function of $\hat{r}$ and ϕ (Gordon & Kogelnik 2000)

$$\vec{\tau} = \phi_w \hat{r} + \hat{r}_w \sin\phi + \hat{r}_w \times \hat{r}(\cos\phi - 1) \tag{34}$$

where $\phi = \Delta\beta\Delta z$ represents the rotation angle of the polarization state vector $\hat{s}$ around the birefringence vector $\vec{\beta}$, and ϕ_w and $\hat{r}_w$ represent their first derivatives of frequency. Eq.(34) obtains that the angle and direction of rotation control the resultant vector $\vec{\tau}$. Substituting the first definition in Eq.(33) into (34), yields

$$\vec{\tau} = \begin{bmatrix} \tau_1 \\ \tau_2 \\ \tau_3 \end{bmatrix} = \begin{bmatrix} \varepsilon\,\Delta z\cos(2\alpha) \\ \varepsilon\,\Delta z\sin(2\alpha) \\ 0 \end{bmatrix} + 2\frac{d\alpha}{dw}\begin{bmatrix} -\sin(\phi)\sin(2\alpha) \\ \sin(\phi)\cos(2\alpha) \\ 1 - \cos(\phi) \end{bmatrix} \tag{35}$$

where $\varepsilon = d\Delta\beta / dw$ represents PMD parameter, and Δz is the fiber segment length. On the other hand, $\vec{\tau}$ is a function of w , which may be written as a Taylor series around the central frequency w_o as follows (Agrawal 2005)

$$\vec{\tau}(w) = \vec{\tau}(w_o) + \Delta w\frac{d\vec{\tau}}{dw}|_{w=w_o} + \frac{\Delta w^2}{2}\frac{d^2\vec{\tau}}{dw^2}|_{w=w_0} + \dots\dots\dots \tag{36}$$

Comparing Eqs.(35) and (36), the first term on the right hand side of Eq.(35) will represent the first order of PMD vector, while the second term indicates all higher orders of PMD vector. Accounting that the higher orders depend on the value of $d\alpha / dw$. For a very small variations of α with frequency, the second term on the right hand side of Eq.(35) may be neglected. Elsewhere, the higher order effects must be included through the determination of PMD vector.

5.1 Linear birefringence

Neglecting the higher order effects makes the PMD vector as follows

$$\vec{\tau} = \begin{bmatrix} \tau_1 \\ \tau_2 \\ \tau_3 \end{bmatrix} = \begin{bmatrix} \varepsilon\,\Delta z\,\cos(2\alpha) \\ \varepsilon\,\Delta z\,\sin(2\alpha) \\ 0 \end{bmatrix} = \varepsilon\,\Delta z\,\hat{r} = \frac{\varepsilon\Delta z}{\Delta\beta}\,\vec{\beta} = \text{const.}\,\vec{\beta} \tag{37}$$

This means, $\vec{\tau}$ coincides with the birefringence vector $\vec{\beta}$ if the intrinsic birefringence is linear and the higher order PMD effects are neglected. Elsewhere, the two vectors are never coincided. Using Eq.(37), we can obtained DGD of the fiber segment as

$$DGD_1 = \tau_{pmd}^{(1)} = |\vec{\tau}| = \varepsilon\,\Delta z \tag{38}$$

The value of DGD_1 represents the delay time between the two components of polarization in a single segment of the optical fiber. Since the DGD's of the fiber segments are random, so that DGD_1 can be calculated as $<\tau_{pmd}>= \dfrac{1}{N}\sum_{i=1}^{N}\tau_{pmd}^i$. For the case of wide frequency band, the higher order effects of the PMD must be included. The DGD_2 of this case can be obtained using Eq.(35) as follows

$$DGD_2 = \tau_{pmd}^{(2)} = \sqrt{(\varepsilon\,\Delta z)^2 + 8(1-\cos\phi)^2\alpha_w^2} \tag{39}$$

Clearly, the DGD_2 is related to the change of α with respect to frequency, and $\tau_{pmd}^{(1)} < \tau_{pmd}^{(2)}$. This means that the higher order effects increase the DGD. The angle between the two vectors $\vec{\tau}$ and $\vec{\beta}$ is determined as: $\psi = \cos^{-1}(\tau_{pmd}^{(1)} / \tau_{pmd}^{(2)})$. This means that the two vectors in the same direction if the higher order PMD is neglected, i.e. $\tau_{pmd}^{(1)} =< \tau_{pmd}^{(2)}$.

5.2 Nonlinear birefringence

For the nonlinear intrinsic birefringence, $\vec{\tau}$ can be calculated using the second definition in Eq.(33) and (34) as follows

$$\vec{\tau} = \begin{bmatrix} (a_1 + a_3\sin\phi)\cos(2\alpha) + a_6(\cos\varphi - 1)\sin(2\alpha) \\ (a_1 + a_3\sin\phi)\sin(2\alpha) - a_6(\cos\varphi - 1)\cos(2\alpha) \\ a_2 + a_5\sin\phi \end{bmatrix} + \frac{da}{dw}\begin{bmatrix} -a_4\sin\phi\sin(2\alpha) + a_7(\cos\phi - 1)\cos(2\alpha) \\ a_4\sin\phi\cos(2\alpha) + a_7(\cos\phi - 1)\sin(2\alpha) \\ a_8(\cos\phi - 1) \end{bmatrix} \tag{40}$$

where the parameters a_1 into a_8 are defined as

$$a_1 = \frac{\Delta\beta\varepsilon\Delta z}{\Delta\beta_{NL}} \qquad\qquad a_2 = -\frac{\Delta\beta\varepsilon\;\zeta\;T}{\Delta\beta_{NL}^2}$$

$$a_3 = a_5 = -\frac{\zeta\;T\Delta\beta\varepsilon}{K} \qquad\qquad a_4 = \frac{2\Delta\beta}{\Delta\beta_{NL}}$$

$$a_7 = \frac{\zeta\;T\;\Delta\beta^2\varepsilon}{\Delta\beta_{NL}^2} \qquad\qquad a_8 = -\frac{2b_L}{\Delta\beta_{NL}^2}$$

$$\Delta\beta_{NL} = \sqrt{\Delta\beta^2 + (\zeta\ T)^2} \qquad K = (\Delta\beta^2 + \zeta^2 T^2)^{3/2}$$

$$a_6 = \frac{2\zeta\ T\Delta\beta}{\Delta\beta_{NL}} - \frac{(\zeta\ T)^3\varepsilon}{K}$$

Eq.(40) represents a new formula of the PMD vector demonstrating the difficulties to compensate the noise that arises due to PMD when the pulse propagates through optical fibers. Many approaches have been proposed (McCurdy et al. 2004;Lima et al. 2001; Vanwiggeren & Ray 1999; Ibragimv & Shtenge 2002; Schuh et al. 1995), which deal only with the first order of PMD. This means that the compensation depends on the first term presented in the right hand side of Eq.(40) and assuming that the birefringence vector is linear.

The vector $\vec{\tau}$ can be found from $\vec{\beta}$. Ignoring the higher orders of the vector $\vec{\tau}$, the vector $\vec{\tau}$ is linear only if $\vec{\beta}$ is linear, otherwise they are different. When the distance is changed this implies to rotate $\hat{s}$ around $\vec{\beta}$ by an angle φ. On the other hand, the change of frequency causes to rotate $\hat{s}$ around $\vec{\tau}$ by an angle θ. Fig.(5 a) illustrates the relation among the three vectors $\hat{s}$, $\vec{\beta}$, and $\vec{\tau}$ where the polarization vector $\hat{s}$ is rotating around $\vec{\beta}$ and $\vec{\tau}$. Adding the higher orders of $\vec{\tau}$, the vector $\vec{\tau}$ is now nonlinear which does not coincided with the vector $\vec{\beta}$ as illustrated in Fig.(5 b). The general case considers the birefringence vector is nonlinear and assuming all orders of $\vec{\tau}$ as illustrates in Fig.(5 c), which shows that each vector rotates in Stokes space.

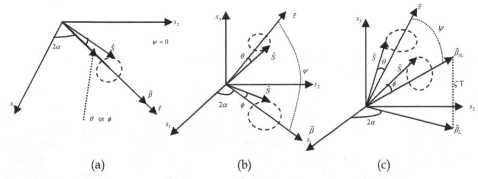

(a)   (b)   (c)

Fig. 5. Rotation of SOP around $\vec{\beta}$ and $\vec{\tau}$: a) $\vec{\beta}$ and $\vec{\tau}$ are linear, b) $\vec{\beta}$ is linear and $\vec{\tau}$ is nonlinear, c) $\vec{\beta}$ and $\vec{\tau}$ are nonlinear.

6. Combined PMD and PDL effects

As far as the continuum limit at the end is set, the following simple arrangement are considered: each PMD element (having $\vec{\tau}_i$ vector) is followed by a PDL element (having $\vec{\alpha}_i$) leading to the following transmission Jones matrix (Yasser 2010)

$$T = T_{PDL}T_{PMD} = \exp(\frac{1}{2}\vec{\alpha}_j.\vec{\sigma})\exp(-\frac{iw}{2}\vec{\tau}_j.\vec{\sigma}) \qquad (41)$$

where

$$\exp(\frac{1}{2}\ \vec{\alpha}_j.\vec{\sigma}) = \ [\cosh(\alpha_j/2) + (\hat{\alpha}_j.\vec{\sigma})\sinh(\alpha_j/2)]$$

$$\exp(-\frac{iw}{2}\vec{\tau}_j.\vec{\sigma}) = \ [\cos(w\tau_{pmd}^{(j)}/2) - i(\hat{p}_j.\vec{\sigma})\sin(w\tau_{pmd}^{(j)}/2)]$$

Here $\vec{\tau}_j = \tau_{pmd}^{(j)}\hat{p}_j$ represents the j-th PMD segment having DGD $\tau_{pmd}^{(j)}$ and the fast polarization axis is expressed by the unit vector $\hat{p}_j$ in the Stokes space. The PMD vector $\vec{\tau}$ is, generally, frequency dependent; the first term in the Taylor expansion of $\vec{\tau}(w)$ is conventionally referred to as the first-order PMD (Agrawal 2005). To clarify the notation used in this section, we attempt to keep the notation simple and transparent while linking to the notation already established as much as possible. The following is an abbreviated group of present notation: The letters C, c, S, and s represent $\cos(w\tau_{pmd}^{(j)}/2)$, $\cosh(\alpha_j/2)$, $\sin(w\tau_{pmd}^{(j)}/2)$, and $\sinh(\alpha_j/2)$, respectively.

Notice that, in this representation PDL matrix, the polarization component of the field that is parallel to $\vec{\alpha}_j$ experiences a gain e^α, but the anti-parallel component is attenuated by $e^{-\alpha}$. The expressions $e^{\pm\alpha_j/2}$ represent the eigenvalues λ_1, λ_2 of PDL matrix. The vector $\vec{\alpha}_j = \alpha_j\hat{\alpha}_j$ stands for the j-th PDL segment with value expressed in dB by

$$PDL(dB) = 10\ell og_{10}(\frac{\lambda_1}{\lambda_2})^2 = 20\,|\,\alpha_j\,|\,\ell og_{10}(e) \tag{42}$$

The action of an optical component exhibiting PDL and PMD on a field can be described by (Chen et al. 2007)

$$|s> = T\,|\,t> = T_{PDL}T_{PMD}\,|\,t> \tag{43}$$

where $|s>$ and $|t>$ are output and input SOP, respectively. The eigenvalues of the matrix $T = T_{PDL}T_{PMD}$ are (Yasser 2010)

$$\lambda = [cC - i(\hat{\alpha}\cdot\hat{p})sS] \pm \sqrt{[cC - i(\hat{\alpha}\cdot\hat{p})sS]^2 - 1} \tag{44}$$

It was evident from Eq.(44) that the eigenvalues are complex, where the real part will control the new rotation angle of $\hat{s}$ around the PSP vector, and imaginary part can be used into Eq.(42) to obtain the PDL value in presence of PMD. Obviously, the new eigenvalues in presence of the combined PMD and PDL effects are different from that obtained for each effect separately.

6.1 Special cases

1. In presenting PDL only, the eigenstates of the PDL matrix are orthogonal, the output Stokes vector can be obtained as follows: combining the relations $|s>=T_{PDL}\,|\,t>$, and $<s|=<t\,|\,T_{PDL}^\dagger$ into $\hat{s} =<s\,|\,\vec{\sigma}\,|\,s>$, and using the facts (Yasser 2010; Gordon & Kogelnik 2000)

$$(\vec{\alpha}\cdot\vec{\sigma})(\vec{\beta}\cdot\vec{\sigma}) = \vec{\alpha}\cdot\vec{\beta} + i\vec{\alpha}\times\vec{\beta}\cdot\vec{\sigma} \tag{45a}$$

$$(\vec{\beta}\cdot\vec{\sigma})(\vec{\alpha}\cdot\vec{\sigma}) = \vec{\alpha}\cdot\vec{\beta} - i\vec{\alpha}\times\vec{\beta}\cdot\vec{\sigma} \tag{45b}$$

$$(\vec{\beta}\cdot\vec{\sigma})\vec{\sigma} = \vec{\beta}\text{-}i\vec{\beta}\times\vec{\sigma} \tag{45c}$$

$$\vec{\sigma}(\vec{\beta}\cdot\vec{\sigma}) = \vec{\beta} + i\vec{\beta}\times\vec{\sigma} \tag{45d}$$

$$(\vec{\beta}\cdot\vec{\sigma})\vec{\sigma}(\vec{\beta}\cdot\vec{\sigma}) = 2\vec{\beta}(\vec{\beta}\cdot\vec{\sigma}) - \beta^2\vec{\sigma} \tag{45e}$$

$$T_{PDL}^{\dagger} = T_{PDL} = c + (\vec{\alpha}\cdot\vec{\sigma})s \tag{45f}$$

a useful relation can be deduced

$$\hat{s} = c^2\hat{t} - s^2\hat{t} + 2sc\hat{\alpha} + 2s^2\hat{\alpha}(\hat{\alpha}\cdot\hat{t}) \tag{46}$$

The output SOP which is a combination of the vectors $\hat{t}$ and $\hat{\alpha}$, i.e. $\hat{t}$ does not rotate around $\hat{\alpha}$. If the input SOP is parallel or anti-parallel to PDL then the output SOP takes the form $e^{-\alpha}\hat{t}$ or $e^{\alpha}\hat{t}$. The first component, that is parallel to PDL vector, experience a gain e^{α} and the other, that is anti-parallel to PDL vector, is attenuated by $e^{-\alpha}$.

2. Similarly, in presence of PMD only, the eigenstates of the PMD matrix are also orthogonal and the output SOP can be determined as follows: combining the relations $<s|=<t|T_{PMD}^{\dagger}$ and $|s> = T_{PMD}|s>$ into $\hat{s} =<s|\vec{\sigma}|s>$, using Eqs.(45) with the facts that $T_{PMD} = C - i(\hat{p}\cdot\vec{\sigma})S$ and $T_{PMD}^{\dagger} = T_{PMD}^{-1} = C + i(\hat{p}\cdot\vec{\sigma})S$ to yield

$$\hat{s} = C^2\hat{t} - S^2\hat{t} + 2SC(\hat{p}\times\hat{t}) + 2S^2\hat{p}(\hat{p}\cdot\hat{t}) \tag{47}$$

This equation refers to the input SOP that are parallel or anti-parallel to PMD vector which experiences no change, i.e. $\hat{s} = \hat{t}$ along the optical fiber. Notice that, the PMD causes a rotation of the SOP around $\vec{\tau}$, which is presented through the third term.

3. Finally, in presenting the combined PDL-PMD effects, determining $\hat{s}$ as a function of $\hat{t}$, $\hat{\alpha}$, and $\hat{p}$ which is very complicated, is beyond the scope of this chapter.

6.2 The output power

The normalized Gaussian pulse before entering the PMD and PDL components has the form

$$|A_{in}(T) > = D\,e^{-T^2/2T_o^2}\,|a> \tag{48}$$

where T_o is the initial pulse width, and $|a>$ is the Jones vector of the signal. Clearly, the normalized input power is found to be $\vec{P}_{in}(T) =< A_{in}(T)|\vec{\sigma}|A_{in}(T) > = e^{-T^2/T_o^2}\hat{s}$, where $\hat{s}$ is the input Stokes vector. The Fourier transform of Eq.(48) is

$$|A_{in}(w) > = \Im\{|A_{in}(T) >\} = D\frac{T_o}{\sqrt{2\pi}}e^{-w^2T_o^2/2}|a> \tag{49}$$

As far $|A_{out}(w) > = T_{PDL}T_{PMD}(w)|A_{in}(w) >$, the output field which can be illustrated by the inverse Fourier transformation as follows

$$|A_{out}(T) > = D\frac{T_o}{\sqrt{2\pi}}T_{PDL}\Im^{-}\{e^{-w^2T_o^2/2}e^{-i(w/2)(\vec{\tau}\cdot\vec{\sigma})}\}|a> = e^{-\frac{T^2+\tau_{pmd}^2}{2T_o^2}}e^{\vec{\alpha}\cdot\vec{\sigma}}e^{-T\vec{\tau}\cdot\vec{\sigma}/2T_o^2}|a> \tag{50}$$

In order to compute the output power from this equation. The vector $\vec{n}$ was set to equal $\vec{n} = (\vec{\alpha} - \vec{\tau}\,T/T_o^2)/2$, such that

$$| A_{out}(T) >= e^{-\frac{T^2+\tau_{pmd}^2}{2T_o^2}} e^{\vec{n}\cdot\vec{\sigma}} | a >$$

(51)

The new vector $\vec{n}$ is a random. Its value is $n = \sqrt{\alpha^2 + T^2\tau_{pmd}^2 / T_o^4 - 2T\alpha\tau_{pmd}^2 \cos\theta / T_o^2} / 2$, where θ is the angle between $\vec{\alpha}$ and $\vec{\tau}$, while the direction is $\hat{n} = (\vec{\alpha} - T\vec{\tau}/T_o^2)/n$. Substituting Eq.(51) into the definition $\vec{P}_{out}(T) = < A_{out} | \vec{\sigma} | A_{out} >$ and introducing the fact $(\vec{n}\cdot\vec{\sigma})^\dagger = \vec{n}\cdot\vec{\sigma}$, yields

$$\vec{P}_{out}(T) = e^{-\frac{T^2+\tau_{pmd}^2}{T_o^2}} < a | (\cosh n + \hat{n}\cdot\vec{\sigma}\sinh n)\vec{\sigma}(\cosh n + \hat{n}\cdot\vec{\sigma}\sinh n) | a >$$

(52)

Considering Eqs.(45), the last equation may be written as

$$\vec{P}_{out}(t) = e^{-\frac{T^2+\tau_{pmd}^2}{T_o^2}} [\hat{s} + 2\hat{n}(\sinh n \cosh n + \sinh^2 n \cos\phi_1)]$$

(53)

where ϕ_1 is the angle between the random vector $\hat{n}$ and the input SOP, $\hat{s}$. To visualize the situation more easily, Eq.(53) was written as

$$\vec{P}_{out}(T) = e^{-\frac{T^2+\tau_{pmd}^2}{T_o^2}} \cdot f(T,\tau_{pmd},\alpha) \cdot \hat{s}_{out}$$

(54)

where $f(T,\tau_{pmd},\alpha)$ and $\hat{s}_{out}$ are the value and direction of the expression inside the square brackets. Eq.(54) represents the output power in presenting of PMD and PDL, which may be written in certain cases as in the following subsections.

6.2.1 PMD only

In this case, $\vec{n} = -T\vec{\tau}/2T_o^2$ and $\hat{n} = -\hat{p}$, hence, Eq.(53) can be simplified as

$$\vec{P}_{out}(T) = e^{-\frac{T^2+\tau_{pmd}^2}{T_o^2}} [\hat{s} - 2\hat{p}(\sinh n \cosh n + \sinh^2 n \cos\phi_2)] = f(T,\tau_{pmd})e^{-\frac{T^2+\tau_{pmd}^2}{T_o^2}} \hat{s}_{out}$$

(55)

Here ϕ_1 is replaced by ϕ_2 which represents the angle between $\hat{\tau}$ and $\hat{s}$. If $\vec{\tau} = 0$, then $\vec{P}_{in} = \vec{P}_{out}$. That is; the power and SOP are not affected in absence of PMD. The PSP's are the states that are parallel or antiparallel to $\hat{p}$, so the powers in the PSP's direction are $\vec{P}_{out}(T)_{PSP} = \exp(-(T^2 \pm \tau_{pmd}^2)/T_o^2)\hat{s}$. The parallel or antiparallel SOP to $\hat{p}$ will not be changed through the propagation, but the position of the pulse components will be shifted by $\pm\tau_{pmd}/2$.

6.2.2 PDL only

Here, $\vec{n} = \vec{\alpha}/2$ and $\hat{n} = \hat{\alpha}$, hence, Eq.(53) will be

$$\vec{P}_{out}(T) = e^{-\frac{T^2}{T_o^2}} [\hat{s} + 2\hat{\alpha}(\sinh n \cosh n + \sinh^2 n \cos\phi_3)] = f(\alpha)e^{-\frac{T^2}{T_o^2}} \hat{s}_{out}$$

(56)

Here ϕ_1 is replaced by ϕ_3 which represents the angle between $\hat{\alpha}$ and $\hat{s}$. If $\vec{\alpha} = 0$, then $\vec{P}_{in} = \vec{P}_{out}$. That is; the power and SOP are not affected in absence of PDL. There are two important SOP's that are parallel or antiparallel to $\hat{\alpha}$. For these SOP's, Eq.(56) will be reduced to $\vec{P}_{out}(T) = e^{\pm \alpha} e^{-T^2/T_o^2} \hat{s}$. This means that, the power will be affected by the factor $e^{\pm \alpha}$ but the pulse shape and SOP will not change.

6.3 The complex PSP vector

Before discussing the impact of PMD and PDL on the dynamical equation of SOP, we notice: First, without including PDL, the transmission matrix of the fiber is always unitary. However, when the fiber PMD is intertwined with PDL elements, the transmission matrix loses its unitary property. Nevertheless, by the polar decomposition theorem (Kogelnik & Jopson 2002), a complex 2×2 matrix can be decomposed into $T = T_{PDL} T_{PMD}$, where T_{PDL} is a positive definite Hermitian matrix, i.e. $T_{PDL}^\dagger = T_{PDL}$, and T_{PMD} is a unitary matrix, i.e. $T_{PMD}^\dagger T_{PMD} = I$. Second, the PDL vector may by frequency dependent. This will influence the PDL induced waveform distortion effect in an optic link. Considering that such frequency dependent waveform distortion is not so important in a system with realistic parameters (Shtaif & Rosenberg 2005; Phua & Ippen 2005), the PDL vector was approximated as a frequency independent.

As pulses are described by wave packets with a finite frequency band, the frequency dependence of $|s>$ should be considered now. A fixed input polarization was assumed, i.e. $|t>_w=0$ hence $\hat{t}_w = 0$, as is appropriate for a pulse entering the fiber at zero time. Now, by differentiating Eq.(43) with respect to frequency and eliminating $|t>$, the change of the output Jones vector was obtained

$$\frac{d|s>}{dw} = T_{PDL} T'_{PMD} T_{PMD}^{-1} T_{PDL}^{-1} |s> \tag{57}$$

where T'_{PMD} represents the derivative of T_{PMD} with respect to frequency. Eq.(57) tell us that for most input polarizations, the output polarization will change with frequency in the first order. Notice that, if $|s>$ either of the two eigenstates of the operator $T_{PDL} T'_{PMD} T_{PMD}^{-1} T_{PDL}^{-1}$ then $|s>_w=0$. The dynamical equation of SOP in Stokes space can be obtained by using Eq.(57) as, see (Yasser 2010)

$$\hat{s}_w = [(c^2 + s^2)\vec{\tau} - 2s^2(\vec{\tau} \cdot \hat{\alpha})\hat{\alpha} + 2isc(\vec{\tau} \times \hat{\alpha})] \times \hat{s} \tag{58}$$

Many published studies (Chen et al. 2007; Wang & Menyuk 2001; Shtaif & Rosenberg 2005) related to the theoretical treatment of the combined effects of PMD and PDL, which are introduced in many forms of the frequency derivative of Stokes vector, but all these forms may be considered as a partial form of Eq.(58) above.

The expression between brackets in the right hand side of the last equation represents the complex PSP vector which can be decomposed as real and imaginary parts as follows

$$\vec{W} = \vec{\Omega} + i\vec{\Lambda} \tag{59}$$

where $\vec{\Omega}$ and $\vec{\Lambda}$ represent the new vectors in presenting of PMD and PDL. The two new vectors take the forms

$$\vec{\Omega} = (c^2 + s^2)\vec{\tau} - 2s^2(\vec{\tau} \cdot \hat{\alpha})\hat{\alpha} \tag{60a}$$

$$\vec{\Lambda} = 2sc(\vec{\tau} \times \hat{\alpha}) \tag{60b}$$

There are many features that can be deduced from Eq.(59): if $\vec{\tau}$ is parallel or anti-parallel to $\hat{\alpha}$ then $\vec{\Omega} = \vec{\tau}$, i.e. the old and new PMD vectors are identical, and $\vec{\Lambda} = 0$, i.e. the PDL effects will disappear. If $\vec{\tau}$ is perpendicular on $\hat{\alpha}$ then $\vec{\Omega} = (c^2 + s^2)\vec{\tau}$, i.e. the old and new PMD vector have the same direction but distinct values, and $\vec{\Lambda} = 2sc\vec{\xi}$ (where $\vec{\xi} = \vec{\tau} \times \hat{\alpha}$) that means the new PDL vector is perpendicular to the plane that contains $\vec{\tau}$ and $\hat{\alpha}$. If $\vec{\tau} = 0$ then both vectors $\vec{\Omega}$ and $\vec{\Lambda}$ are zero. Remembering that, the absence of PMD will not permit the emergence of two components, as a result there is no PDL but the reverse is not correct. Since the PSP vector is complex, then the fast and slow PSP's are not orthogonal. If $\hat{\alpha} = 0$, i.e. no PDL, then $\vec{\Omega} = \vec{\tau}$. The new DGD takes the form $\tau_{pmd}^{new} = \mathrm{Re}\sqrt{\vec{W} \cdot \vec{W}} = \tau_{pmd}^{old}$, where the meaning of DGD over infinite frequency is called the scalar PMD. Thereafter, the SOP rotates around the PSP vector by an angle $\tau_{pmd}^{new}w$. The new DAS takes the form $\alpha_{new} = \mathrm{Im}\sqrt{\vec{W} \cdot \vec{W}}$. Accordingly, the new PDL value is $20\,|\,\alpha_{new}\,|\,\ell og_{10}(e)$.

7. Birefringence and nonlinearity

To formulate the birefringence effects more precisely, considering the nonlinear Helmholtz equation (Agrawal 2007)

$$\nabla^2 \vec{\tilde{E}} + \frac{w^2 \vec{\tilde{\varepsilon}}_s}{c^2} \vec{\tilde{E}} = -\frac{w^2}{c^2 \varepsilon_o} \vec{\tilde{P}}_{NL} \qquad (61)$$

where the tilde denotes the Fourier transformation, ε_o is the vacuum permittivity, and $\vec{\tilde{\varepsilon}}_s$ is the linear part of the dielectric constant. Notice that the tensorial nature is important to account for the PMD effects that have their origin in the birefringence of silica fibers, while its frequency dependence leads to chromatic dispersion. Assuming that the instantaneous electronic response dominates and neglecting Raman contribution (Lin & Agrawal 2003 a), the third order nonlinear polarization in a medium as silica glass is found to be

$$\vec{\tilde{P}}_{NL}(w) = \frac{\varepsilon_o \chi_{xxxx}^{(3)}}{4}\left[(\vec{\tilde{E}} \cdot \vec{\tilde{E}})\vec{\tilde{E}}^* + 2(\vec{\tilde{E}}^* \cdot \vec{\tilde{E}})\vec{\tilde{E}}\right] \qquad (62)$$

The electric field vector evolves along the fiber length and its SOP changes because of the birefringence. It is assumed here that the z-axis is directed along the fiber length and The electric field vector lies in the x-y plane. This assumption amounts to neglect the longitudinal component of the vector and is justified in practice as long as the spatial size of the fiber mode is longer than the optical wavelength. In Jones-matrix notation, the field at any point r inside the fiber can be written as (Kogelnik & Jopson 2002)

$$\vec{\tilde{E}}(r, w) = F(x, y)\,|\,A(z, w) > e^{ikz} \qquad (63)$$

where $F(x, y)$ represents the fiber mode profile, k is the propagation constant, and Jones vector $|A>$ is a two-dimensional column vector representing the two components of the electric field in the x-y plane. Since $F(x, y)$ does not change with z, one needs to consider only the evolution of $|A>$ along the fiber.

Substituting Eq.(63) into Eq.(62), inserting the result into Eq.(61), and integrate over the transverse mode distribution in the x-y plane, assuming $|A>$ to be slowly varying function of z so that neglecting their second-order derivative with respect to z. With these simplifications, the equation governing the evolution of $|A>$ takes the form

$$\frac{d|A>}{dz}+\left(\frac{w^2\ddot{\varepsilon}_s}{2ikc^2}+i\frac{k\sigma_o}{2}\right)|A>=\frac{i\gamma}{3}\Big[2<A|A>+|A^*><A^*|\Big]|A> \tag{64}$$

where σ_o is a unit matrix. To proceed Eq.(64) further, the dielectric constant tensor $\ddot{\varepsilon}_s$ may be represented in the basis of Pauli matrices as (Lin & Agrawal 2003 a)

$$\frac{w^2\ddot{\varepsilon}_s}{c^2}=\left[k+i\frac{\alpha}{2}\right]^2\sigma_o-k\vec{\beta}.\vec{\sigma} \tag{65}$$

The vector $\vec{\beta}$ accounts for the fiber birefringence and its frequency dependence produces PMD. The vector $\vec{\sigma}$ is formed as $\vec{\sigma}=\hat{e}_1\sigma_1+\hat{e}_2\sigma_2+\hat{e}_3\sigma_3$, where $\hat{e}_1$, $\hat{e}_2$, and $\hat{e}_3$ are a three unit vectors in the Stokes space. Substituting Eq.(65) into (64) leads to the following vector equation

$$\frac{d|A>}{dz}+\frac{\alpha}{2}\sigma_o|A>=-\frac{i}{2}\vec{\beta}.\vec{\sigma}|A>+\frac{i\gamma}{3}\Big[2<A|A>+|A^*><A^*|\Big]|A> \tag{66}$$

Eq.(66) can be put in simplified form by neglecting the second term on the left hand side, by proposing that the medium is lossless; then, using the following identity

$$|A^*><A^*| = [<A|A>+<A|\vec{\sigma}|A>\cdot\vec{\sigma}]/2-<A|\sigma_3|A>\sigma_3 \tag{67}$$

into Eq.(66) yields the following elegant equation that describes the evolution of Jones vector through the optical fiber

$$\frac{d|A>}{dz}=(-\frac{i}{2}\vec{\beta}.\vec{\sigma}+\frac{i\gamma}{6}[\hat{s}.\vec{\sigma}]^t)|A> \tag{68}$$

where the proportionality term $|A>$ affects only the global phase and can be neglected, $\hat{s}=<A|\vec{\sigma}|A>$ is the normalized power (Stokes vector). Using Eq.(68), it is not difficult to obtain

$$\frac{d\hat{s}}{dz}=(\vec{\beta}+2\gamma(0,0,s_3)^t/3)\times\hat{s} \tag{69}$$

Eq.(69) presents the effect of nonlinearity. Introducing γ effect is considered as the main contribution of this section, because it is a phenomenon that can not be neglected in the study of the evolution of polarization through the optical fibers. However, the rotation axis in presence of nonlinearity is $\vec{\beta}+2\gamma(0,0,s_3)^t/3$ instead of $\vec{\beta}$. The simplest case, without nonlinearity effect, has been studied by many researches using different approaches, see for example (Gordon & Kogelnik 2000; Agrawal 2005; Vanwiggeren and Roy 1999).

8. Conclusions

In conclusion, we have achieved the following: an important mathematical relationship between PMD and birefringence are presented and all possible assumptions are discussed. The statistics of PMD are simply analyzed. The combined effect of PMD and chromatic dispersion causes an additional amount of pulse broadening. Interaction of PMD and PDL makes the two PSP's are not orthogonal nor do they represent the fastest and slowest pulses, which causes a change in DGD and PDL compared with the impact of each individual. Nonlinearity causes a change in the rotation axis and therefore it changes the properties of polarization state during the propagation. Finally, all results are generally subject to random changes as long as most of the causes random.

9. References

Agrawal G. P. (2007). *Nonlinear Fiber Optics*, 4th Edition, Academic Press, USA.

Agrawal G. P. (2005). *Lightwave Technology: Telecommunication Systems*, 1st Edition, Wiley Interscience, USA.

Azzam R. M. and Bashara N. M. (1989). *Ellipsometry and Polarized Light*, Elsevier, Amsterdam.

Chen L., Hadjifaradji S., Waddy D., and Baw X. (2003). Effect of Local PMD and PDL Directional on the SOP Vector Autocorrelation, *Optics Express*, Vol.11, No.23, pp.3141-3146.

Chen L., Zhang Z., and Bao X. (2007). Combined PMD-PDL Effects on BERs in Simplified Optical System: an Analytical Approach, *Optics Express*, Vol.15, No.5, pp.2106-2119.

Chertkov M., Gabitov I., Kolokolov I., and Schafer T. (2004). Periodic Compensation of Polarization Mode Dispersion, *J. Opt. Soc. Am. B*, Vol.21, No.3, pp.486-497.

Elbers J., Glingener C., Duser M., and Voges E. (1997). Modeling of Polarization Mode Dispersion in Single Mode Fibers, *Elect. Let.*, Vol.33, No.22, pp. 662-664.

Foshchini G. and Poole C. (1991). Statistical Theory of Polarization Dispersion in Single Mode Fibers", *J. Lightwave Tech.*, Vol.9, pp.1439-1456.

Gordon J. and Kogelnik H. (2000). PMD Fundamentals: Polarization Mode Dispersion in Optical Fibers", *Proc. Natl. Acad. Sci.*, Vol.97, No.9, pp.4541-4550.

Ibragimv E. and Shtenge G. (2002). Statistical Correlation Between First and Second Order PMD, *J. Lightwave Tech.*, 20(4): 586-590.

Karlsson M. (1994). Polarization Mode Dispersion Induced Pulse Broadening in Optical Fibers", *Optics Let.*, 23, pp.688-690.

Kogelnik H. and Jopson R. M. (2002). Polarization Mode Dispersion, *in Optical Fiber Telecommunications volume: IV B, I. P. Kaminov and T. Li, Eds. San Diego: Academic*, pp.725-861, USA.

Levent A., Rajeev S., Yaman F., and Agrawal G. P. (2003). Nonlinear Theory of Polarization Mode Dispersion for Fiber Solitons, *Phys. Rev. Let.*, Vol.90, No.1, pp.730-737.

Lima T., Khosravani R., and Menyuk C. R. (2001). Comparison of PMD Emulators", *J. Lightwave Tech.*, Vol.19, No. 12, pp.1872-1881, 2001.

Lin Q. and Agrawal G. P. (2003), Vector Theory of Stimulated Raman Scattering and Its Application to Fiber-based Raman Amplifier, *J. Opt. Soc. Am. B.*, Vol.20, No.8, pp.492-501.

Lin Q. and Agrawal G. P. (2003). Correlation Theory of Polarization Mode Dispersion", *J. Opt. Soc. Am. B*, Vol.20, No.2, pp-292-301.

Lin Q. and Agrawal G. P. (2003). Statistics of Polarization Dependent Gain in Fiber Based Raman Amplifiers, *Optics Let.*, Vol.28, No.4, pp.227-229,.

Mahgerftech D. and Menyuk C.R. (1999). Effects of First-Order PMD Compensation on the Statistics of Pulse Broadening in Fiber with Random Varying Birefringence, *IEEE Photo. Tech. Lett.* 13(3): 340-342.

McCurdy A., Sengupta A., and Glodis P. (2004). Compact Measurement of Low PMD Optical Telecommunication Fibers, *Optics Express*, Vol.12, No.6, pp.1109-1118.

Mechels S., Schlger J., and Franzen D. (1997). Accurate Measurements of the Zero Dispersion Wavelength in Optical Fibers, *J. Res. Natl. Inst. Stand. Tech.*, Vol.102, No.3, pp.333-347.

Menyuk C.R., Wang D., and Pilipetskii A. (1997). Re-polarization of Polarization Scrambled Optical Signals Due to PDL, *IEEE Pho. Tech. Let.*, Vol.9, No.9, pp.1247-1249.

Phua P. and Ippen E. (2005). A Deterministic Broad Band Polarization Dependent Loss Compensator", *J. Lightwave Tech.*, Vol.23, No. 2, pp.771-780.

Rogers A. (2008). *Polarization in Optical fibers*, Artech House, INC, USA.

Schuh R., Sikora E., Walker N., Siddiqui A., Gleeson L., and Bebbington D. (1995). Theoretical Analysis and Measurements of Fiber Twist on the Differential Group Delay of Optical Fibers, Electron Let., Vol.31, No.20, pp.1772-1773.

Shtaif M. and Rosenberg O. (2005). Polarization Dependent Loss as a Waveform Distortion Mechanism and Its Effect on Fiber Optic Systems, *J. Lightwave Tech.*, Vol.23, No. 2, pp.923-930.

Sunnerud H., Karlsson M., Xie C., and Andrekson P. (2002). Polarization Mode Dispersion in High Speed Fiber Optic Transmission Systems", J. Lightwave Tech., Vol.20, No.12, pp.2204-2219.

Tan Y., Yang J., Kath W., and Menyuk C. (2002). Transient Evolution of Polarization Dispersion Vector's Probability Distribution, *J. Opt. Soc. Am. B*, Vol.19, No.5, pp.992-1000.

Vanwiggeren G. and Roy R. (1999). Transmission of Linearly Polarized Light through a Single Mode Fiber with Random Fluctuations of Birefringence, *Applied Optics*, Vol.38, No.18, pp.3888-3892.

Wang D. and Menyuk C. (2001). Calculation of Penalties Due to Polarization Effects in Long-Haul WDM System Using a Stokes Parameter Mode, *J. Lightwave Tech.*, Vol.19, No. 4, pp.487-494.

Yaman F., Lin Q., Radic S., and Agrawal G. (2006). Fiber Optic Parametric Amplifiers in the Presence of Polarization Mode Dispersion and Polarization Dependent Loss, *J. Lightwave Tech.*, Vol.24, No. 8, pp.3088-3096.

Yang J., Kath W., and Menyuk C. (2001). Polarization Mode Dispersion Probability Distribution for Arbitrary Distances, *Optics Let.*, Vol.26, No.19, pp.1472-1474.

Yasser H. A. (2010). The Dynamics of State of Polarization in the Presence of Conventional Polarization Effects, *Optics & Laser Tech.*, 42, 1266-1268.

Yoon I. and Lee B. (2004). Change in PMD Due to the Combined Effects of PMD and PDL for a Chirped Gaussian Pulse, *Optics Express*, Vol.12, No.3, pp.492-501.

8

Spun Fibres for Compensation of PMD: Theory and Characterization

Lynda Cherbi and Abderrahmane Bellil
Laboratory of Instrumentation LINS,
University of Sciences and Technology Houary Boumedienne, USTHB,
Algiers,
Algeria

1. Introduction

The polarization is a relative property to the vibratory nature of light. In an optical fibre, light is a combination of two vibrations of perpendicular directions. Each direction represents one mode of polarization. Indeed, the optical fibres and the components of the optical fibres present a small difference in the refractive index in the pair of the polarization states, a property called the birefringence. This last one induces a difference of propagation speed between the two modes. So, light at the output, cannot be restored more faithfully. The birefringence can change the state of polarization (SOP) of light when it crosses the fibre. In a single mode fibre, the birefringence is combined with a random coupling of polarization modes. The delay measured at the output of the fibre between the two polarization modes is called the difference of group delay DGD (measured in picoseconds). The polarization modes dispersion (PMD) results from the variation of the DGD according to the wavelength and the environment conditions.

The typical tolerance of a system to the PMD is roughly 10% of the bit period, which gives 40 Ps for a system of 2.5 Gb/s, 10 Ps for a system of 10 Gb/s and only 2.5 Ps for a system of 40 Gbs/s (Noé et al.,1999). The PMD is a random phenomenon and constitutes an enormous obstacle ahead of the increase of the debits from 10 Gbit/s for a part of the networks of most telecommunication companies. Several solutions have been proposed to compensate the PMD as: The electronic compensation after a direct photo-detection that can only eliminate a part of the PMD effects since the information about the polarization and the phase get lost at the detection; the second solution is the electronic compensation in a coherent receptor with diversity of polarizations, and the third one is the optical compensation in at least a differential delay section. Other solutions are proposed by the Corning society and which rely on the use of spun fibres allowing the control of the coupling of the modes, therefore reducing the PMD; thus, giving differential group delays of order of Femtoseconds. In the past decade, some considerable efforts have been made to understand the origins of the PMD and to attenuate its effects in the systems. The PMD can be reduced in a fibre with two different manners. The first one consists in minimizing the asymmetries in the refractive index profile and the constraints, which implies improvements of the industrial process in the manufacture of the fibre in order to assure a better geometry and to reduce the rate of constraints in the fiber. The second method allows the control of the coupling of the modes of the polarization in the fibre while spinning it during its manufacture. Indeed, the spinning has been used in the manufacture of fibres since the beginning of the 1990s, and it showed that it is an efficient technique to reduce the PMD in the fibre. First, we start by presenting in this work the spun fibres explaining their technology, their principle and their different types. Next, the description of the reduction of the PMD by using the spinning is developed by a mathematical formalism based on the theory of coupling and Jones's matrix. Moreover, the reduction of the PMD is verified in the spun fibres while applying the method of JME and the COTDR method (photon counting -Optical temporal Domain of Reflectometry) that allowed us to measure the DGD of the order of femtosecond (Cherbi et al., 2009). The comparison of the DGD found in this type of new generation of fibres with those of the standard ones, led us to confirm that the spun fibres offer effectively a smaller DGD than those of the standard fibres, emphasizing the importance of this type of fibres in the reduction of the PMD.

We present the different results already published (Cherbi et al., 2009) while using the reflectometers COTDR and POFDR (polarization- Optical Frequency Domain of Reflectometry) which are used to get the polarization characteristics of the spun fibres as the beat length and the PMD and to observe the spatial frequencies linked directly to the period of spinning.

2. Principle of spun fibres

2.1 Technologies of the spun fibres
There are more than two decades when the concept of the spun fibres has been proposed originally in an article published by (Barlow et al.,1981).

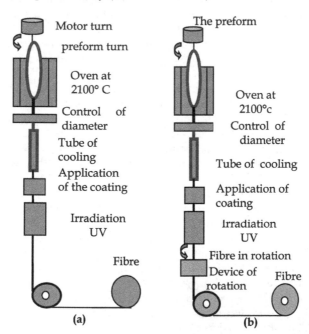

Fig. 1. Two approaches used to present the rotating fibres (a) turn preforms (b) turn the fibre.

The fibre spun is achieved by the rotation of the preform during the drawing of the fibre (figure 1.a). In this approach, the system of drawing is the same as that of the conventional standard fibre systems (OVD) except that a rotating motor is placed on the top of the preform. When the motor is set in motion with a predetermined speed, the preform starts turning dragging the rotations of the axes of the birefringence. The rotation will end up with the end of the drawing operation. This approach is quite simple and appropriate for the pulling of the fibre at low speed. However, this is not convenient for the production of the fibre with a high speed pulling because the rotation of the motor must be at very high speed as well. To illustrate this, we consider a rotation rate of the fibre of 3 turns/min, for a drawing speed of 1 m/s, the rotating speed of the preform is thus only 180 turns/min.

On the other hand, for a modern drawing device having a speed higher than 20m/s, the perform must turn at a speed greater than 3600 turns/mn, which is far from practical. For this reason the concept of the spun fibres has not been used in the production of fibres until the half of the nineties when methods of more adapted spinning have been proposed (Ming-Jun & Nolan, 1998). Moreover, the transmission systems as they appeared at low rate (<= 2.5 Gb/s), the PMD was not a major problem to seek fibres that perform this reduction.

Several convenient techniques have been suggested during the year 1990, for example, by (Hart et al., 1994] in order to make the fibre turn rather than the preform. Later on, this technique became the most adapted one for the manufacturing of the fibres performing the reduction of the PMD.

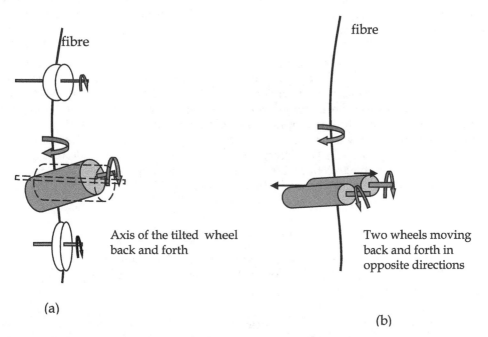

Fig. 2. Examples of fibre rotation systems: (a) tilted wheel, (b) two wheels moving in opposite directions.

In this case (fig 1.b), a rotating device of the fibre, is put along the way of fibrage to rotate the fibre directly. Two examples of this device are illustrated in figue 2. In the first example (fig 2.a), a wheel is in contact with the fibre and tilted with respect to its initial position, thus applying a moment of rotation to turn the fibre. In the second example (fig 2.b), two wheels are placed horizontally and are both in contact with the fibre (Blaszyk & Christoffand,2001). The two wheels move back and forth in opposite directions driving the rotational movement of the fibre. Imposing a direct motion to the fibre eliminates the problem of the preform when turning at high speed. Besides, this technique provides flexibility to control and implement different profiles of rotation for a better reduction of the PMD.

2.2 Theory of spun fibres

Two approaches have been suggested to model the reduction of the PMD in the spun fibres, one of which is based on the evolution of the polarization state (Galtarossa et al., 2001; Ming-Jun &Nolan, 1998). The evolution of the vector representing the polarization dispersion is ruled by the dynamic equation which is linked to the vector of the local birefringence. While solving the dynamic equation, the vector representing the polarization's dispersion is gotten and its module gives the delay of the differential group (DGD). Another approach is based on the theory of the coupled modes of Jones' matrix (Ming-Jun et al., 2002) where the complex amplitudes of the two modes of polarization are described by the equations of the modes coupling. While solving these equations, the complex amplitudes are derived and Jones' matrix is determined and the DGD can be computed from this matrix. Basically, the two approaches give equivalent results. Our survey of the spun fibres is founded on Jones' matrix formalism, where we notice that the analytical solutions obtained are simple.

2.2.1 Equations of the mode coupling

As the birefringence in the fibres used in telecommunications is generally small, the formalism based on the theory of disruption (Ming-Jun et al., 2002), can be used to describe the different mechanisms of birefringence in the single mode fibres, including the birefringence due to the distortion of the core, constraints, curvature, rotation of the fibre and torsion. In what follows, we will present the theory of the coupled modes and we will show how to implement it in the different problems of birefringence. Indeed, the small birefringence of telecommunication fibres can be treated as an anisotropic disruption to a material originally isotrope. In the condition of weak guidance, the electric field E is described by the following wave equation (Dandliker,1992):

$$\Delta E - \mu_0\, \varepsilon_0\, \varepsilon\, E = \mu_0 p \tag{1}$$

Where ε_0 et μ_0 are the dielectric and magnetic constants of vacuum respectively, ε is the relative dielectric constant of the non disrupted fibre, and p is the disruption term given by the following relation:

$$p = \varepsilon_0\, \Delta\varepsilon\, E \tag{2}$$

Where $\Delta\varepsilon$ is the electric tensor describing the anisotropy of the medium. Without the term of disruption, the equation (1) has modal solutions of the following shape:

$$E_n(x,y,z) = e_n(x,y)\exp(-i\beta_0 z) \qquad n = 1,2 \tag{3}$$

where $e_n(x,y)$ is the distribution of the electric field. For a monomode fibre, n =1, 2 represent the two modes of polarization.

In absence of disruption, the two modes are degenerated and propagate with the same constant β_0. In presence of the disruption term, it is supposed that the electric field E(x,y,z) is given by the linear superposition of the two non disrupted modes (Ming-Jun et al., 2002):

$$E(x,y,z) = \sum_n A_n(z)e_n(x,y)\exp(-i\beta_0 z) \tag{4}$$

Where $A_n(z)$ are the complex coefficients describing the amplitudes and the phases of the two modes. Let's put equation (4) into equations (1) and (2) and use the relation of orthogonality between the two modes (Ming-Jun et al., 2002):

$$\int e_m(x,y) \cdot e_n(x,y)dxdy = \begin{cases} N_m & m = n \\ 0 & m \neq n \end{cases} \tag{5}$$

Knowing that N_m is a constant of normalization which can be calculated as follows:

$$N_m = \tfrac{1}{2}\int \vec{e}_m \times \vec{h}_m^* \, z \, ds = \frac{n_{coeur}}{2}\left(\frac{\varepsilon_0}{\mu_0}\right)^{1/2} \int e_m^2 ds \tag{6}$$

and using the condition of the weak coupling:

$$\frac{1}{\beta_0}\left|\frac{d^2 A_n}{dz^2}\right| << \left|\frac{dA_n}{dz}\right| \tag{7}$$

We get the equations of the coupled modes that describe the evolution of the complex amplitudes $A_n(z)$:

$$\frac{dA}{dz} = ik.A \tag{8}$$

where A is the complex amplitude vector taking the following form:

$$A = (A_1 \quad A_2)^T \tag{9}$$

and k is the matrix of the coupling coefficients.

$$k = \begin{pmatrix} k_{11} & k_{12} \\ k_{21} & k_{22} \end{pmatrix} \tag{10}$$

The coupling coefficients are associated to the different types of disruptions:

$$k_{mn} = \frac{k_0}{2n_0N_0} \int e_n^*(x,y).\Delta\varepsilon(x,y,z). e_m(x,y)dxdy \tag{11}$$

Where n_0 is the effective refractive index of both non disrupted modes.

2.2.2 Jones matrix and the PMD of spun fibre

The evolution of the local polarization along the birefringent fibre is described by the equations of the modes coupling. The total change of polarization of an input signal, after having traveled a given distance in the fibre is better described by Jones' matrix. Let's assume that the losses in the fibres are negligible, the already predefined Jones matrix, can be put under another form which is:

$$T = \begin{bmatrix} A_1(z) & -A_2^*(z) \\ A_2(z) & A_1^*(z) \end{bmatrix} \text{ with } |A_1|^2 + |A_2|^2 = 1 \tag{12}$$

The four complex elements of Jones' matrix can be gotten while integrating the equations of the coupled modes with suitable initial conditions. Once Jones' matrix is known, the PMD can be calculated easily from the elements of the matrix (Chen, 2002; Ming-Jun et al., 2002):

$$\tau = 2\sqrt{\left|\frac{dA_1}{d\omega}\right|^2 + \left|\frac{dA_2}{d\omega}\right|^2} \tag{13}$$

In order to describe the reduction of the PMD, we define a parameter, named reduction factor of PMD (PMDRF) ζ as the ratio of the DGD of the spun fibres over the DGD of the standard fibre.

$$\zeta = \frac{\tau}{\tau_0} \tag{14}$$

Where the used lengths for the spun fibres and standard fibres are the same. For example, if ζ is equal to 1, the reduction of the PMD is not achieved and if ζ is equal to 0.5, a factor of two is obtained in the reduction of the PMD.

2.3 Different types of spun fibres

The coupling coefficients matrix depends upon the dielectric tensor of the disruption. The values of these elements are determined by the type of disruption, which means that they depend on the configuration of the fibre. In this section, we describe some configurations of the fibres and we give their coupling coefficients matrix. It is important to note that the coupling matrixes in this work are expressed on the basis of the circular polarization because it is more appropriate to process the rotating fibres (Ming-Jun et al., 2002).

2.3.1 The linearly birefringent fibre

The linear birefringence is a consequence of disruptions as the distortions of the core, the asymmetry of the lateral constraints, the curvature. In the case of the linear birefringence, the coupling coefficients matrix is given by (Ming-Jun et al., 2002):

$$k = \frac{1}{2} \begin{pmatrix} 0 & \Delta\beta e^{i2\varphi} \\ \Delta\beta e^{-i2\varphi} & 0 \end{pmatrix} \qquad (15)$$

where $\Delta\beta$ is the linear birefringence, and Φ is the orientation of the birefringence with respect to a given axis

2.3.2 Spun fibres
In a spun fiber, the orientation of the birefringence takes place depending on the x axis. The rotation angle Φ accumulated is therefore a function of the fibre length 'z', which in turn is determined by the rate of rotation $\alpha(z)$:

$$\Phi = \int_0^z \alpha(z)\, dz \qquad (16)$$

Replacing the equation (16) in equation (15), we get the coupling coefficients matrix of the rotating fibres, describing the disruption of the birefringence,:

$$k = \frac{1}{2} \begin{bmatrix} 0 & \Delta\beta e^{i2\int_0^z \alpha(z)dz} \\ \Delta\beta e^{-i2\int_0^z \alpha(z)dz} & 0 \end{bmatrix} \qquad (17)$$

2.3.3 Twisted fibre
There are two effects in this type of fibres: The rotation of the birefringence and the mechanical torsion. The rotation of the birefringence is similar to that of the rotating fibre. If the rate of torsion is T, the angle Φ is calculated by

$$\Phi = T\, z \qquad (18)$$

The rate of torsion is determined by the coefficients of photo - elasticity of the fibre. The torsion constraint induces the circular birefringence proportionally to the rate of torsion.

$$\delta = g.T \qquad (19)$$

Where g is the coefficient determined by the coefficients of photo elasticity of the glass. The typical value of g for fibres in silica is 0.16. Combining both effects of rotation and torsion, the coupling matrix comes up with the following form:

$$k = \frac{1}{2} \begin{bmatrix} \delta & \Delta\beta\, e^{i2Tz} \\ \Delta\beta e^{-iTz} & -\delta \end{bmatrix} \qquad (20)$$

2.4 Solutions of the coupled equations for different types of the spun fibres

Generally, the matrix of the coupling coefficients depends upon the variable z, and the analytic solutions of equation (8) have no existence in the majority of the cases. The numerical integration is always used to get numerical solutions. Different methods, as the method of the finite differences, the Runge-Kutta method (Chen, 2002), can be applied to solve the equation of the coupled modes. However, in the two following special cases, we can derive the analytic solutions which will be discussed in this section.

2.4.1 Constant spinning rate

For a constant spinning rate, the function 'spin' (rotation) can be written as follows:

$$\alpha = \alpha_0 \tag{21}$$

Where α_0 is a constant. In this case, the birefringence of a fibre is estimated in only one direction with a rate α_0. For this reason, the constant spinning rate is often assigned to an unidirectional spinning. For a spun fibre, with a constant spinning rate, the integral of the coupling matrix can be calculated easily, and the coupled equations become (Hart, 1994):

$$\frac{dA_1}{dz} = \frac{1}{2} i \, \Delta\beta \, e^{i2\alpha_0 z} A_2 \tag{22}$$

$$\frac{dA_2}{dz} = \frac{1}{2} i \, \Delta\beta \, e^{i2\alpha_0 z} A_1 \tag{23}$$

With initial conditions $A_1(0) = 1$, $A_2(0) = 0$.

The solutions of equations (22) and (23) are:

$$A_1 = -\frac{\alpha_0 - \upsilon}{2\upsilon} e^{i(\alpha_0 + \upsilon)z} + \frac{\alpha_0 + \upsilon}{2\upsilon} e^{i(\alpha_0 - \upsilon)z} \tag{24}$$

$$A_2 = \frac{\Delta\beta}{4\upsilon} e^{i(-\alpha_0 + \upsilon)z} - \frac{\Delta\beta}{4\upsilon} e^{-i(\alpha_0 + \upsilon)z} \tag{25}$$

Where $\upsilon = \sqrt{\alpha_0^2 + \frac{1}{4}\Delta\beta^2}$

Using equations (12) and (13), we find that the DGD can be expressed by a simple equation for the spinning constant:

$$\tau(z) = \frac{\gamma_\omega}{2\upsilon} \sqrt{(\Delta\beta)^2 z^2 + \left(\frac{4\alpha_0}{\Delta\beta} \sin\left(\frac{\Delta\beta \, z}{2}\right)\right)^2} \tag{26}$$

Where $\gamma_\omega = \frac{d\Delta B}{d\omega} = \frac{\tau_0}{L}$ is the PMD of a uniform birefringent fibre without modes coupling at the z position. The sinusoidal term of equation (26) doesn't play an important role when the fibre is sufficiently long. On the other hand, for long fibres, the DGD is given by:

$$\tau(z) = \frac{\gamma_\omega.\Delta\beta.z}{2\upsilon} \tag{27}$$

Equation (27) indicates that the DGD progresses linearly with the length of the fibre, and the PMDRF takes the following form:

$$\zeta = \frac{\Delta\beta}{2\upsilon} \tag{28}$$

We notice that for spun fibres of constant rate, the PMDRF depends upon the length of beating or the birefringence.

2.4.2 The periodic spin function

For the functions of periodic spin, under some conditions, we can describe analytic solutions by using the theory of disruption (Chen et al., 2002) in which fibres are submitted to uniform disruptions only, or in the case of small lengths regime (typically smaller than 100 m) in order to fine down their analysis. Indeed, in this approach, the random characteristic of the variation of the disruption in case of important lengths regime is ignored. Using the initial conditions issued from the previous paragraph, the first order solutions of disruption for A1 (z) and A2(z) are as follows:

$$A_1(z) = 1 \tag{29}$$

$$A_2(z) = (i/2)\Delta\beta \int_0^z \exp[-2i\,\Theta(z')]dz' \tag{30}$$

Where $\Theta(z) = \int_0^z \alpha(z')dz'$

It becomes easier to obtain the DGD by using equation (13):

$$\tau(z) = \gamma_\omega \left| \int_0^z \exp[-2i\,\Theta(z')']dz' \right| \tag{31}$$

Based on the theory of disruption, the first order of the disruption's expansion is valid only when $\Delta\beta \ll 1$. This condition puts some limits on the application of equation (31) on fibres that have a low PMD.

The validity of this solution has been tested by (chen et al., 2002). When the length of beating is important (some meters), i.e. $\Delta\beta \approx 1$, and the period of spin is smaller than the length of beating, the theory of disruption of the first order can always be applied. For sinusoidal profiles of spin, the expression for the factor of the PMD reduction can be gotten from the solutions of the disruption equation. Let's notice that the profile of sinusoidal spin takes the following form:

$$\alpha(z) = \alpha_0 \cos(\eta z) \tag{32}$$

Where α0 is the spin amplitude , and η is the angular frequency of the spatial modulations, which is linked to the spin period Λ through the following relationship $\eta = 2\pi/\Lambda$.

With the analytical solution of equation (31), we are able to assert which spin parameters give right to the optimization of the PMD performances. With the first observations, we remark that, when the length of beating of a fibre is bigger than some meters, the PMDRF is independent of the beating length, and therefore of the intrinsic birefringence of the fibre. In equation (31), the only contribution to the birefringence of the fibre, comes from γ_ω, and the DGD is proportional to this size. Let's note that γ_ω is the PMD of the unspun fibers (non rotating). On the other hand, the PMDFR will be independent of γ_ω.

This conclusion is also verified by the direct numerical integration of equation (8) with k given by equation (17). Some old fibres had beating lengths inferior to some meters; with the improvement of fibre manufacturing, the majority of these lengths were improved lately beyond some meters. The PMDRF independence from the intrinsic birefringence of the fibre, offers the advantage of simplicity in its conception because it is worthless to optimize the spin profiles for the different birefringences of the fibre. Moreover, we noticed that the DGD increases linearly when the length of the fibre increases (figure 3) despite the fact that we got some overlapping oscillations on the graph representing the variation of the DGD with respect to the distance.

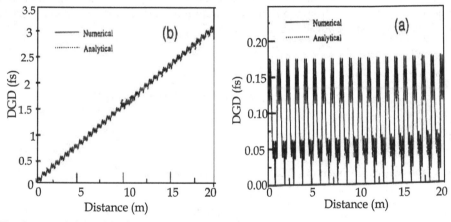

Fig. 3. Evolution of the DGD along a spun fibre.

We can also separate the real and imaginary contributions of the integral (31) in order to better analyze the variation of the DGD along the fibre. We express the equation (31) in an alternative way of the DGD for one spatial period T:

$$DGD(T) = \gamma_\omega \left| \int_0^T \cos[2\Theta(z')]dz' - i\int_0^T \sin[2\Theta(z')]dz' \right| \tag{33}$$

$$DGD(T) = \gamma_\omega \int_0^T \cos[2\Theta(z')]dz', \quad \Theta(z) \text{ is an even function}$$

$$DGD(T) = \gamma_\omega \int_0^T \sin[2\Theta(z')]dz', \quad \Theta(z) \text{ is an odd function}$$

We suppose that $\Theta(z)$ is a periodic function. When $\Phi(z)$ is an even function, $\int_0^T \sin[2\Theta(z')] dz'$ is equal to zero. When $\Theta(z)$ is an odd function, $\int_0^T \cos[2\Theta(z')] dz'$ is equal to zero. For multiple values of the period T, the DGD becomes $n[DGD(T)]$. For values in between, some oscillations are encrusted in the linear variation of the DGD. On the other hand, this survey based on [40] leads us to the conclusion that the dependence of the DGD of the standard fibres on the square root of their lengths, comes from the statistical nature of the random coupling of the two modes of polarization. The linear evolution of the DGD with respect to the length of the spun fibre is caused by the periodicity of the coupling induced by the spinning, thus we have a coupling mode better-controlled than in the case of standard fibres. However, it is possible that the DGD of the spun fibres follows a different evolution law in a region where the first order theory of disruption is not valid any more; for example, when the intrinsic birefringence of the fibre is high and / or the spin rate is high.

With the aforementioned results, it is rather simple to find the phase matching conditions for which the maximum reduction of the PMD can be obtained. In this case, the condition is fixed such that the PMDFR is equal to zero (chen et al., 2002):

$$\int_0^T \exp[-2i\Theta(z')] dz' = 0 \qquad (34)$$

Equation (34) can be expressed in another way if we use the properties discussed previously for even and odd functions. We notice that when the phase matching conditions are satisfied, the evolution of the DGD along the spun fibre is periodic. The DGD doesn't increase anymore when the length of the fibre increases.

Equations (31) and (34) are valid for a whole category of periodic profiles of spin. To illustrate the way how to determine the phase matching conditions, we take an example of a sinusoidal spin profile. Such a profile is defined by equation (32). The integration of this profile gives $\Theta(z) = \alpha_0 \sin(\eta z) / \eta$; then we get the DGD by using equation (31):

$$DGD(z) = \gamma_\omega \left| \int_0^z \exp[-i\frac{2\alpha_0 \sin(\eta z')}{\eta}] dz' \right| \qquad (35)$$

The integral can be valued analytically by using the following identity:

$$\exp[-ix\sin(\theta)] = J_0(x) + 2\sum_{n=1}^{\infty} J_{2n}(x).\cos(2n\theta) - 2i\sum_0^{\infty} J_{2n+1}(x)\sin[(2n+1)\theta] \qquad (36)$$

Then, we get

$$DGD(z) = \gamma_\omega \left[R^2(z) + I^2(z) \right]^{1/2} \qquad (37)$$

where

$$R(z) = J_0(2\alpha_0 / \eta)z + \sum_{n=1}^{\infty} \frac{J_{2n}(2\alpha_0 / \eta)}{\eta n} \sin(2n\eta z) \qquad (38)$$

$$I(z) = \sum_{n=0}^{\infty} \frac{J_{2n+1}(2\alpha_0 / \eta)}{\eta(2n+1)} \cos[(2n+1)\eta z] \qquad (39)$$

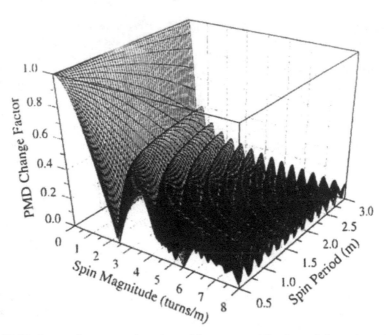

Fig. 4. The PMD change factor as a function of the spin amplitude and the spin period.

We note that when $J_0(2\alpha_0 / \eta) \neq 0$, the dominant contribution comes from the term of the linear increase of equation (38). Neglecting the oscillations term, the expression of the DGD becomes:

$$DGD(z) = \gamma_\omega J_0(2\alpha_0 / \eta)z \qquad (40)$$

As in the case of spun fibres with a constant rate, in absence of random disruptions, the DGD increases linearly with the length of the fibre; in contrast with the PMDRF which takes a simpler shape:

$$PMDRF = J_0(2\alpha_0 / \eta) \qquad (41)$$

Equation (41) indicates that the PMDRF is independent from the beating length in the case of spun fibres with sinusoidal profile whose beating lengths are equal to some meters or more. When $J_0(2\alpha_0 / \eta) = 0$, the linear increase term disappears, and the oscillation terms cannot be neglected any more. In this case, the DGD oscillates between 0 and a maximum value and is independent of the propagating distance. The condition where the minimum of the PMD is reached is called the condition of phase matching (figure 4).

Figure 4 illustrates the presentation in three dimensions for the graph of the PMD reduction as a function of the spin period and the spin amplitude. The phase matching condition can be achieved for different spin parameters enabling to get an optimal reduction of PMD; though in general, amplitudes of higher spins give a better reduction of PMD. Figure (3.a) shows the evolution of DGD along the fibre with the phase matching condition. Finally, the maximum of reduction can be reached at the zeroes of the Bessel function of order zero (equation 41).

2.5 Reduction of the PMD for different profiles of spin
2.5.1 The constant spin rate
While using equation (28), the PMDRF, as function of the spin rate, is represented for different beating lengths in figure 5; for a constant spin rate.

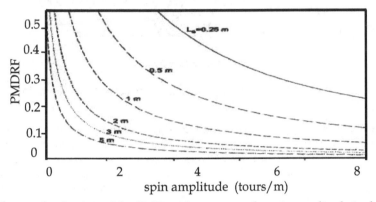

Fig. 5. The factor of reduction of the PMD with respect to the spin amplitude in the case of a constant spinning rate for different beat lengths.

We note that the PMD is reduced when the spin rate increases. For the same spin rate, PMDRF depends on the beating length. The higher the beating length is, the more reduced is the PMD. For a high PMD of the fibre (beating length <1m), a high spin rate is necessary to reduce the PMD.

2.5.2 Sinusoidal spin
In figure 6, we use a beating length of 1m as example to illustrate the reduction of the PMD for sinusoidal types of spin (Ming-Jun et al., 2002).
Figure 6 shows that for sinusoidal spin types, the PMDRF oscillates with the spin amplitude, which is different from the case where the spin was constant. Furthermore, this figure shows that, for a sinusoidal spin, the phase matching condition can be gotten in order to come to a low PMD; on the other hand, in the case of constant spin, the phase matching doesn't exist. The phenomenon of phase matching can be explained by the mechanism of coupling of modes. The constant spin reduces the birefringence of the fibre, and causes no coupling of modes as well. For the sinusoidal spin, the variation in the rate of spin carries along the two modes of polarization to intercouple, reaching a compensation of the PMD. For some spin profile and birefringence of fibre, the conditions of phase matching are satisfied and the maximum of energy exchange occurs in order to provide a better reduction of PMD. The results of modeling indicate that the conditions of phase matching depend on the beating length, the period of the spin and the amplitude of the spin. We can use the same function of

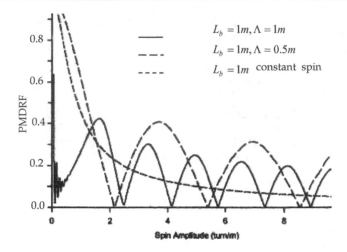

Fig. 6. The factor of reduction of the PMD versus the spin amplitude in the sinusoidal spin profiles.

spin to get a small reduction of PMD for high lengths of beating. However, for small lengths of beating; the phase matching has a strong dependence with the length of beating.

The fact that the birefringence of the real fibres is not constant and changes randomly, it is impossible to have the phase matching for the whole birefringence while using only one sinusoidal spin. This problem can be solved by admitting spin profiles with many Fourier components. To get to this point, the concept of the use of the modulated spin in amplitude and frequency has been developed by the Corning society.

2.6 Statistical evolution of the PMD of the spun fibres

As it was mentioned in the previous sections, the spun fibres follow a linear evolution law without the random modes coupling or in the régime of short lengths. When the random mode coupling is present, it has been found that the spun fibres follow an evolution law, a function of square root, similar to that of the unspun fibres, but with a different rate depending on the spin parameters (Chen, 2002). The random mode coupling can be characterized by a random variation of the birefringence axis and / or by the induced phase shift by the external constraints with an occurrence frequency of $1/h$, where h is called the coupling length of the modes. On the other side, a fibre of length «l» can be divided into (l/h) segments. Using this model, for a sinusoidal profile of spun fibre under no optimal conditions (no phase matching), the DGD can be expressed under the following simple form:

$$\tau = \zeta \gamma_\omega \sqrt{hl} \tag{42}$$

We notice that the fact that the PMDRF «ζ» is independent of the beating length when the length of beating is greater than some meters, the DGD in the régime of important lengths, and in presence of the random coupling mode, is corrected by a factor ζ, which is the reduction induced by the fibre spinning during the process of drawing. In this case, the property of evolution of the PMD is similar to that of the fibre possessing the linear birefringence.

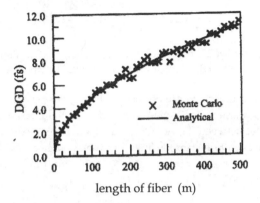

length of fiber (m)

Fig. 7. The DGD of a spun fibre according to the length of the fibre. The amplitude of spin is 3.5 turn/m, the period of spin is 1m, the length of beating is 10m and the length of coupling is 10m.

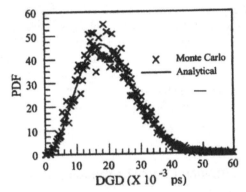

Fig. 8. Probability density function versus the DGD of the fibre when the condition of phase matching is not verified. The amplitude of spin is 3.5 turn/m, the period of spin is 1m, the length of fibre is 500m and the length of coupling is 10m.

The simplest law of evolution given by equation (42) has been verified by using a numerical modeling (Chen, 2003). Figure 7 shows the results of the numerical simulation for a sinusoidal spun fibre under the non optimal conditions. As it is shown in this figure, the numerical modeling accomodates very well with the theoretical prediction.

In the case of standard fibres (unspun fibres) with a random coupling mode length h, the distribution of the DGD is analog to the distribution of Maxwell, where the standard deviation σ used in the expression of PDF can be given by the following expression:

$$\sigma = (\frac{\lambda}{cL_b})\sqrt{h.l} / \sqrt{3} \tag{43}$$

We have proven that the Maxwell distribution is valid in the case of the spun fibres, except that, the parameter σ should be corrected by the contribution of the spinning fibre. The modified parameter σ is now under the following form

$$\sigma = \left[J_0(2\alpha_0 / \eta) \right] (\frac{\lambda}{cL_b}) \sqrt{h.l} / \sqrt{3} \tag{44}$$

This equation has been tested and validated in (chen, 2002). Figure 8 represents the probability density function 'PDF ', according to the DGD of the spun fibre, obtained by numerical calculations and Maxwell distribution equation where we confirm according to the figures that the two results converge perfectly.

When the conditions of phase matching are satisfied, the total DGD of the fibre is a periodic function, and it oscillates between the zero value and a maximal value ε_{max}. For this reason, the DGD of only one segment of a fibre is linked to the average of DGD inside one period of spin. Therefore, in the regime of high lengths ($l >> h$), the total DGD can be written as follows (Ming-Jun et al., 2002):

$$\tau = \varepsilon' \varepsilon_q \sqrt{l / h} \tag{45}$$

Where ε_q is the square average of the DGD in one period of spin, and ε' is a coefficient that depends on the average coupling coefficient between two segments. For a condition of phase matching (for example: $\alpha_0 = 2.76 \, tours / m \quad et \, \eta = 2\pi m^{-1}$), ε' is found equal to 1.194. Besides, the DGD increases when the length of coupling of modes decreases (Ming-Jun et al., 1998). It is foreseeable, because under the conditions of phase matching, the DGD is minimum. Any disruption moves the fibre away from the optimal conditions, implying an increase of the PMD. Despite the fact that the DGD of the optimized spun fibres changes differently with the coupling length in comparison with the DGD of the non optimized spun fibres, the DGD always follows a Maxwell distribution, but with a modified parameter σ (Chen, 2002).

$$\sigma = (\varepsilon' \varepsilon_q \sqrt{l / h}) / \sqrt{3} \tag{46}$$

3. Application of the JME method for the measurement of the PMD of the spun fibres

We used the JME method (Derickson, 1998) in order to verify the reduction of the PMD in the spun fibres (Cherbi et al., 2006). This applied method, between 1510 to 1615 nanometers, consists in determining the DGD directly between the two main states of polarization by measuring the Jones matrix of the device under test to a set of wavelengths. In order to determine the PMD of the spun fibres, we take the following steps:

- Measure Jones' matrixes $JM(\lambda_i)$ for a set of wavelengths $\lambda_1, \lambda_2, ... \lambda_n$ of the work range (1510 nm-1615nm)
- Do the product $JM(\lambda_i + \Delta\lambda).JM^{-1}(\lambda_i - \Delta\lambda)$
- Determine the eigen values ρ_1 and ρ_2 of the calculated product of matrix
- The DGD (λ_i) is gotten then by (Heffner, 1992):

$$DGD(\lambda_i) = \Delta\tau = \left| \frac{\arg\left(\frac{\rho_1}{\rho_2}\right)}{\Delta\omega} \right| \tag{47}$$

The PMD of the fibre under test is determined by the arithmetic mean of the $'n'$ measured DGD:

$$PMD = \frac{\sum_{i=0}^{n} DGD(\lambda_i)}{n}$$

We applied the above procedure to two types of the spun fibres in order to compare their performances. The first fibre is unidirectional of length 212m in which we noted that the rotation of the spins was only in one sense once removed it from the spool. On the other hand, for the second bi-directional of length 1Km, the rotation of the spins was in the two senses. The results gotten in the figure 9.a and the figure 9.b, show that this method has a good resolution because it permitted to measure DGD of the order of femtoseconds, and to show that this type of fibre presents effectively low DGD compared to those measured in standard fibres that are of order of the picoseconds. Besides, we noted that the bi-directional fibre possesses a lower DGD than that of the unidirectional one indicating thus the efficiency of the bi-directional spun fibres in the reduction of the PMD.

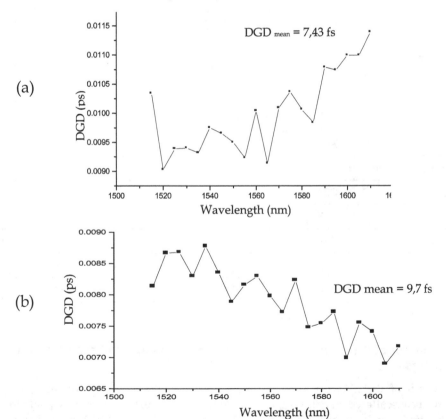

Fig. 9. Representation of DGD measured of spun fibre according to the wavelength for the length (a) L = 290m, (b) L = 212m.

The researchers and inventors of the optical fibre of telecommunication systems predict the impact of PMD from the distribution of $\Delta \tau$, because this results from the variation of $\Delta \tau$ as a function of wavelength and the conditions of the environment. On the other hand, due to this type of variation, the PMD of an optical path is expressed then statistically, as either the average or the root mean square (RMS) of $\Delta \tau(\lambda_i)$ (Derickson, 1998). It is interesting to determine the total PMD of a link made of a series of different spun fibres.

For this reason, we took three different lengths of spun fibres (fibre2, 3 and 4). We started by measuring, with the JME method, their PMD separately, for a given temperature while using a reference fibre (fibre1) used in calibration (table 1). Then, we connected the three fibres, and done the measurement of the total PMD in the same experimental conditions. The same procedure has been applied for the two fibres (fibres 3 and 4). We sought for the best relation of computation to determine the total PMD of a link of spun fibres, by testing the two following relations:

$$PMD_{totale} = PMD_1 + PMD_2 + ...PMD_n \tag{48}$$

$$\text{Or} \quad PMD_{totale} = \sqrt{PMD_1^2 + PMD_2^2 + ...PMD_n^2} \tag{49}$$

With n the number of fibres used in the link

Our experimental results regrouped in table 1 are in very good agreement with the first relation [Cherbi et al., 2006].

	DGD(fs) measured with the step (10nm)	Total DGD (fs) calculated with relation (48)	Total DGD (fs) calculated with relation (49)
reference fibre of (1 km)	98,721		
Fibre 2 (212 m)	4,8223		
Fibre 3 (290m)	7,4315		
Fibre 4 (1 km)	9,7399		
connected Fibres (2+3+4)	22,9437	21,99	13,16
connected fibres (3+4)	17,1985	17,17	12,25

Table 1. The PMD relation of the spun fibres link.

4. Determination of the polarization's properties of the spun fibres using the reflectometers

The beat length of the fibre can be measured directly by the extraction of the spatial period of the backscattered signals (Wegmuller, 2002, 2004), which permits to estimate the PMD in the single-mode fibres (Ellison et al., 1998; Chen, 2002). The OFDR method is not exploited again especially for investigating of the spun fibres for the determination of its parameters and of their PMD according to the distance. In this section, we will present the relation already demonstrated experimentally by COTDR in our anterior works (Cherbi et al., 2009) existing between the spatial period of the backscattered signal and the PMD of the spun fibre and given by (Chen, 2003). Even more, the COTDR method allowed us to compare the results found with those of the JME method. Afterward we will present the POFDR method which used for spun fibres (Cherbi et al., 2009; Wegmuller et al., 2005) to obtain the beat lengths of the two types of spun fibres and the spin period of the bi-directional fibre.

Chen(Chen, 2003) has demonstrated that the spatial period of the backscattered signals obtained from a POTDR (polarization-sensitive optical time-domain reflectometer) of the spun fibres varies linearly with the beat length of the fibre. This means that for a given beat length, the spatial period T_s can be used as calibration for the reduction of the PMD. A simple relation linking the spatial frequency F ($F = 1 / T_s$) to the beat length and the spin parameters, is given by:

$$F = \left| J_0 \left(2\alpha_0 / \eta \right) \right| / (L_b / 2) \tag{50}$$

The PMD of the bi-directional spun fibre is linked to the spatial frequency in the form (Chen, 2003):

$$PMD = \left(\lambda / 2c \right) F \tag{51}$$

Thus, the PMD of the bi-directional spun fibres can be determined directly through the measure of the spatial period as in the case of the standard fibres, while measuring the spatial period of the backscattered obtained from the reflectometers. The equation (50) shows that when the spin is zero, the spatial period converges to the one of the standard fibres.

4.1 Measure of the DGD in the spun fibres by the C-OTDR method

The technique (COTDR) (Wegmuller et al., 2004) is appropriate to detect the defaults in a given fibre (sites of reflection, losses) with a spatial resolution of the decimetre order. The main difference of this reflectometer (Cherbi et al., 2009) compared to a classic OTDR (Ellison & Siddiqui, 1998) resides in the use of photon counting detector (InGaAs avalanche photodiode). It is used in the so-called gated Geiger mode, which means that the detector is only active during a short time slot. During this period, only a single photon falls in the detector and triggers an avalanche, which is then detected by electronics discriminator. Contrary to the operation of a classic detector APD in linear regime, this avalanche is no longer proportional to optical input signal power, but independent of it. The detection is therefore a binary one, either there is an avalanche, or not. In order to evaluate the incident optical power (or mean photon number) on the detector during its activation, the detection process (gate opening) must be repeated many times in order to determine the detection probability of photons with a good precision. This probability is proportional to the incident signal power that is smaller than about 40% (no detector saturation) and larger than the detector thermal noise (dark counts). This condition is satisfied in our set-up by using the variable attenuator before the excitation of the fibre.

From the detection probability for a certain gate position, set by the delay generator, the reflectivity at the corresponding location in the fibre is readily gotten with a spatial 2- point resolution determined by gate duration. Thus, to have some information on the different positions in the fibre, the gate delay must be adjusted. In our set-up, the user can specify the zoom interval (L_{start}, L_{stop}) for which the reflectivity is automatically measured with a step size (sampling resolution).

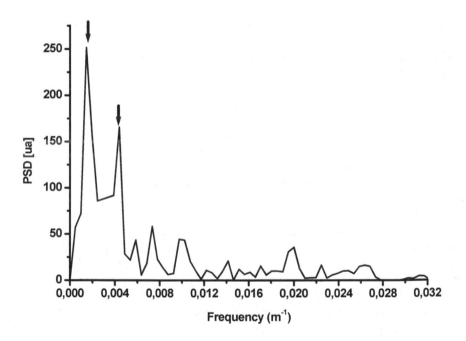

Fig. 10. The Power spectral density of C-OTDR signal for the bi-directional spun fibre of 1 Km.

Figure 10 presents the PSD of the backscattered signal power measured by the C-OTDR reflectometer with its spatial frequencies for a bi-directional spun fibre of 1 Km length. Two peaks appear, respectively, at spatial frequencies F and F/2, in the COTDR trace. A spatial frequency of 0.005 m^{-1} is gotten from the backscattered signal PSD. The relation (51) gives a

DGD of the used spun fibre equal to 11.4 femtosecondes. We found for the same fibre a mean value of the DGD equal to 11,53 femtosecondes while using the JME method. These two results are in very good agreement. In conclusion, this demonstrates that the DGD of a spun fibre can be calculated from the spatial frequency of the COTDR signal in accordance with relation (51) permitting to calculate the spatial frequency of a backscattered signal COTDR in a spun fibre from its parameters": $\alpha_0 = 3.5\pi\ rad$, $\Lambda = 1.5\ m$ and $L_B = 20\ m$. The calculated spatial frequency is equal to $F = 0.005\ m^{-1}$, which is equal to the same one measured from the C-OTDR trace. Based on that, we validate the equation, linking the spatial frequency, the spin parameters and the intrinsic birefringence of the bi-directional spun fibre, given in (Chen, 2003)

4.2 Measure of the beat length of the spun fibres by P-OFDR

This reflectometer implements the technique of coherent detection sensitive to the polarization in order to get information about the evolution of the polarization states along the fibre under test. In our case, a POFDR is used, implementing the detection of polarization diversity (Cherbi et al, 2009] and a polarized beam splitter which plays the role of a fixed analyser. The former permits to remove the Rayleigh reflections independent of the polarization by subtracting output 1 from output 2, thereby removing the frequencies of the back scattered signal that are not related to the fibre birefringence.

The used laser in this reflectometer is a DFB (distributed feedback) characterised by a spectral width of the order of 1MHz on the whole tuning range, a spatial range of 80 m. Due to the coherent detection, a very good sensitivity of 100 dB is gotten with this reflectometer. The only factor limiting the resolution of this method is the tuning of the laser. The laser that we used is limited by the continuous tuning of 20 GHz that gives approximately a resolution of 9 mm.

In (cherbi et al., 2009), we have analyzed three types of fibres having the same length of 200 m: a bi-directional spun fibre, a unidirectional spun fibre and a standard fibre. They were wrapped on a table in order to minimize the external constraints. Figure 11 shows the example of the different POFDR traces for different used resolutions of the unidirectional spun fibre (dark line is the mean of different traces). The beat lengths of the two types of spun fibres and the one of the standard fibre are calculated by the following relation (Wegmuller et al., 2002):

$$< L_b >= \frac{1}{std(DSP)} \sqrt{\frac{12}{\pi}} \qquad (52)$$

Where PSD is the power spectral density of POFDR signal.

The calculated values of beat lengths derived for the PSD signals of the different fibres: unidirectional spun fibre (figure 11), standard fibre and bi-directional spun fibre (figure 12) are respectively: 50 m, 38 m, and 150 m. We note that the beat length of the bi-directional spun fibre is more important than those of the others, which means that the PMD of the bi-directional spun fibre is lower than that of the two other types of fibres, result that we found with the JME method. It also confirms that the bi-directional spun fibre reduce efficiently the PMD compared to the unidirectional spun and the standard fibres.

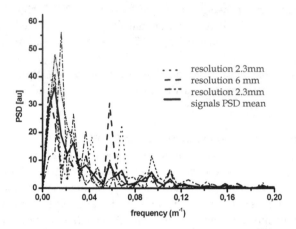

Fig. 11. The mean power spectral density of the backscattered signals POFDR, obtained for different resolutions, of the unidirectional spun fibre of 200 m length.

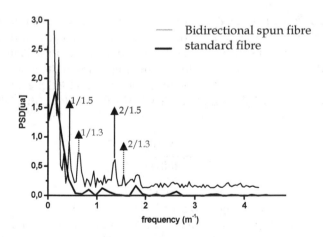

Fig. 12. The mean power spectral density of the backscattered signals POFDR obtained for bi-directional spun fibres and standards of lengths 200m.

5. Conclusion

In this chapter, we presented the principle of the spun fibres with their technology and their role in the reduction of the PMD in a transmission link of optical fibre. Several types of spun fibres have been given. The theory of these fibres based on the equations of the coupled modes has been detailed.

The reduction of the PMD in these fibres is verified while applying the JME method and the COTDR method used to measure the DGD of the order of femtoseconds. We also confirmed this result while measuring the beat length of these fibres with the POFDR method and compare it with that of a standard fibre. This comparison proved the efficiency of this type of fibres in the reduction of the PMD. Finally, according to the use of the COTDR and POFDR, we concluded that the validity of the data analysis, obtained from the reflectometers and used nowadays for standard fibres, has been demonstrated for the spun fibres and more precisely for the bi-directional spun fibres. Besides, the high spatial resolution of the POFDR enables again the observation of the spatial frequencies directly linked to the spin period, so a precise characterization of the spun fibres can be accomplished.

6. References

Barlow, A.J; Ramskov-Hansen; Payne,D.N. birefringence and polarization mode dispersion in spun single mode fibers", *applied optics*, vol.20 (1981), pp 2963.

Blaszyk, P.E. & Christoffand, W.R. method and apparatus for introducing controlled spin in optical fibers", *U.S patent 63224872 B1* (December 2001).

Chen, X.; Ming, J. Li. & Nolan, D. A. Polarization mode dispersion of spun fibers : an analytical solution . *optics Letters*, vol.27(2002),pp. 294 –296.

Chen, X. Scaling properties of polarization mode dispersion of spun fibers in the presence of random mode coupling. *optics letters*, vol. 27 (2002),pp. 1595.

Chen,X. properties of polarization evolution in spun fibers, optics letters, vol. 28 (2003), pp. 2028.

Cherbi,L.; Azrar,A.; Mehenni,M. &Aksas, R. Characterization of the Polarization in the Spun Fibers. *Microwave and Optical Technology Letters*, Vol.51, N°.2, (February 2009), pp. 341–347, ISSN 0895-2477

Cherbi, L.; Mehenni, M. & Wegmuller, M. Mesure de la dispersion des modes de la polarisation dans les fibres optiques spun par la méthode d'analyse des valeurs propres de la matrice de Jones . *25 iémes Journées Nationales d'Optique Guidées JNOG*, pp 176-178, Metz, France, Novembre 2006.

Dandliker, R (1992). Rotational effects of polarization in optical fibers in anisotropic and nonlinear optical waveguides, *C, G. Someda, Elsevier, New york*, pp. 39 – 76.

Derickson. D (1998). *Fiber optic, test and measurement.* Hp professionals books, Prentice hall.

Ellison, G. & Siddiqui, A. S. A fully polarimetric optical time domain reflectometer", *IEEE Photonics technology letters*, vol.10 (1998), pp. 246 -248.

Galtarossa, A.; Palmieri, L. & Pizzinat, A. optimized spinning design for low PMD fibers: an analytical approach. *Journal of lightwave technology*, vol. 19 (2001), pp. 1502.

Hart, A.; GHuf, R. & Walker. method of making a fiber having low polarization mode dispersion due to a permanent spin", *U.S patent 5*, Vol. 47 (1994), pp. 298, March 29.

Heffner,B. Automated measurement of polarization mode dispersion using Jones matrix eigenanalysis. *IEEE Photonics technology Letters*, vol.4 (1992), n°9, pp.1066.

Ming-Jun,Li &Nolan, D. fiber spin profile designs for producing fibers with low polarization mode dispersion. *optics letters*, vol. 23(1998), pp. 1659 – 1661.

Ming-Jun, li; Chen, X. & Nolas, A. fiber spinning for reducing Polarization mode dispersion in single mode fibers: theory and applications, science and technology division, *Corning Incorporated*, Sp – AR 02-2 (2002), Corning NY14831.

Noé et al. Polarization mode dispersion compensation at 10, 20, and 40 Gb/s with various optical equalizers. *IEEE journal of lightwave Technology* , vol. 17 (1999), pp 1602 – 1610.

Wegmuller, M.; Scholder, F. & Gisin,N. Photon counting OTDR for local birefringence and fault analysis in the metro environment. *Journal of lightwave technology*, vol. 22 (2004), N°. 2.

Wegmuller, M.; Legré, M. & Gisin, N. Distributed Beatlength Measurement in single mode fibers with optical frequency domain reflectometry, *journal of light wave technology*, vol 20 (2002), N° 5, pp. 829.

Wegmuller, M.; Cherbi,L. & Gisin, N. Investigation of spun fibers using high resolution reflectometry. *Proceeding OFC IEEE of Optical Fiber Conference*, America 2005.

9

Optical Fibers in Phase Space: A Theoretical Framework

Ana Leonor Rivera
Centro de Fisica Aplicada y Tecnologia Avanzada,
Universidad Nacional Autonoma de Mexico
Mexico

1. Introduction

Since their discovery, optical fibers have received increasing attention due to its important technological applications (Bottacchi, 2002; Culshaw, 1997; Harmon, 2001; Herrmann, 1973; Keyl, 2002; Lauterborn et al., 1997; Prasad & Williams, 1991; Shimizu et al., 1997; Way, 1998; Young, 2000). A fiber is an optical waveguide in which the propagation of an optical wave is confined to two dimensions (the cross section dimensions). The dimension of the confinement must be comparable to the wavelength of the light which one would like to confine (Kogelnik, 1979).

A fiber is formed by a region with a refractive index larger than that of the surrounding media; this condition assures the total internal reflections at the interfaces required for the beam propagation. If an integer number of wavelengths have traveled between two consecutive reflections, a standing wave pattern will be developed, producing a constructive interference which allows high electromagnetic power to be transmitted along the fiber. A change in the fiber geometry, modifies the reflection angle and consequently the total number of reflections: rays traveling thicker fibers require more reflections to travel along the same length in the propagation direction. This implies that the effective velocity of light in thicker fibers must be slower; this effect also produces a change in the phase of the output signal (Arnaud, 1976). These phenomena are cumbersome to explain by standard optical methods (Kogelnik, 1979; Torchigin & Torchigin, 2003) and are easily predicted by the use of the Wigner Distribution Function (Reyes et al., 1999). One of the main advantages of the phase space approach to solve optical problems is the simplicity of the mathematical calculations compared with the traditional treatment (Bottacchi, 2002). This is due to the fact that in phase space representation, the relevant properties of the system can be obtained by simple matrices products.

Quantum mechanically, the development of nonlinear optics allowed the generation and manipulation of new quantum states of light, going from the simplest and common one, the so-called coherent states (Glauber, 1963), to squeezed states (Walls, 1983; Yuen, 1976), Fock states (Lvovsky et al., 2001) or entangled states (Ou et al., 1992). A full quantum theoretical analysis of the three-photon states is contained in the Wigner function (Leonhardt, 2001) that has proven to be very helpful to visualize in the phase space (the amplitude q and phase p quadratures) quantum mechanical system defined by its density matrix. This has already been the case for some quantum states of light such as the coherent state, the squeezed vacuum or the bright squeezed state (Breitenbach et al., 1995; 1997; Smithey et al., 1993), whose Wigner function has been experimentally reconstructed using homodyne quantum tomography, a technique that allows the measurement of the marginal probability distribution that expresses the quadrature amplitude distribution. The use of optical fiber for quantum squeezing has considerable technological advantages, such as generating squeezing directly at the communications wavelength and the use of existing transmission technology (Corney et al., 2008).

More generally, the Wigner function contains the full information about the quantum states (Wigner, 1932) and their moments allow to differentiate between paraxial regime, wave-like regime and chaotic behavior (Rivera et al., 1997). More particularly, it allows us to establish the quantum correlations between the different generated modes in the case of twin photons or photon triplets (Benchekh et al., 2007). The Wigner function is a positive definite function in the phase space only for classical states with Gaussian marginal probability distributions (Rivera & Castano, 2010). However, it can be negative in some circumstances for particular quantum states of light. These negativities are the signature of highly nonclassical behaviour of a quantum state (Rivera & Castano, 2010) as it has been observed for a quantum state of light prepared in a single-photon Fock state (Lvovsky et al., 2001). These quantum negativities are also present in the case of complete degenerate three-photon states obtained by third order optical parametric fluorescence or amplification, and also for aberrated optical systems.

Historically, the so-called *Wigner Distribution Function* (Wigner, 1932) has been of central importance as an alternative description of Quantum Mechanics (Kim & Noz, 1991). However, these phase-space mathematical tool has found exciting applications in a wide range of the physical sciences and even engineering ranging from statistical mechanics (Green, 1951; Mori et al., 1962) to optics (Perinova et al., 1998; Schleich, 2001; Wolf, 2004). Moreover, it has become the basis of an entire discipline: time-frequency representation of wave phenomena (Allen & Mills, 2004; Boashash, 2003; Cohen, 1995; Grochenig, 2000). There exist several reviews of the quantum phase-space distribution functions, in particular of the Wigner distribution function. A concise but authorative review of the quantum distribution functions is that by Wigner (Wigner, 1971). A good mathematical treatment of the quantum distribution functions and related operator algebra is given in the book by (Louisell, 1973). Some extensive reviews of the quantum distribution functions are given by (Balazs & Jennings, 1984; Berry, 1977; Filinov et al., 2008; Groot & Suttorp, 1972; Hillery et al., 1984; Lee, 1995; OConnell, 1983; Takabayasi, 1954). Applications of the Wigner distribution function to Optics are reviewed by (Dragoman, 1997; Dodonov, 2002; Mack & Schleich, 2003; Zalevsky & Mendlovic, 1997), and for the particular case of fibers on the works (Bao & Chen, 2011; Benabid & Roberts, 2011; Benchekh et al., 2007; Corney et al., 2008; Leonhardt, 2001; Rivera & Castano, 2010a).

As Wigner functions, the *Lie Algebra*, due to its mathematical simplicity to solve differential equations by numerical integration, has become an important aid for the solution of different problems in classical and quantum mechanics (Bakhturin, 2003; Frank & van Isacker, 1994; Hamermesh, 1962; Jacobson, 1979). A Lie treatment of geometrical optics and aberrations has been developed by (Dragt & Finn, 1976), and it is a new approach to fiber optics (Reyes et al., 1999; Reyes & Castano, 2000) that simplifies the traditional solution of optical problems (Born & Wolf, 1999) to the determination of the corresponding Symplectic Map associated to the optical system, thus reducing the problem to simple matrices products. The Gaussian Symplectic map helps to find the Wigner distribution function of the probability density of an optical fiber, and from it, it is possible to obtain all the physical information required to analyze the fiber (Rivera & Castano, 2010a).

This chapter presents a brief review of the phase-space analysis applied to fiber Optics, using the Wigner Distribution Function. The hope is that it will show the beauty, elegance and usefulness of this mathematical construction. The rest of the chapter is organized as follows. Section 2 gives a short review of phase space representation using the Wigner Distribution Function which describes some of its important properties and its physical interpretation. Section 3 presents the description of the Maxwell equations under paraxial approach (considering parallel rays close to the optical axis of the system) that describe the light propagation in a fiber by a parabolic type equation that is completely equivalent to the quantum system Schrödinger equation for a bidimensional potential-well time-dependent. Section 4 shows an example, analyzing a gaussian beam propagation through a fiber.

2. Phase space representation

The standard formulation of quantum mechanics either in the Schrödinger (Schrodinger, 1946) or in Heisenberg pictures (Heisenberg, 1930) may create an impression that quantum and classical dynamics are completely different (Dirac, 1935). However, there are representations in which quantum dynamics seems to resemble classical statistical mechanics, and where the state of a quantum system is represented by the quasiprobability distribution in phase space of the corresponding classical system (Kim & Noz, 1991). Of course, there are at least two important differences (Hillery et al., 1984):

1. Quasiprobability distributions may take negative values (unlike the true probability distributions).

2. The classical distribution can be localized at a point in phase space, whereas the quantum distribution must always be spread in a finite phase volume, in agreement with uncertainty relations.

Among different quasiprobability distributions Cohen (1995), the Wigner Distribution Function, introduced by Wigner in 1932[1] (Wigner, 1932), is the only one for which the quantum evolution law coincides with the classical one for the case of linear dynamics (Moyal, 1949). The Wigner distribution function is a real valued quasiprobability distribution containing all information available about the system. Its popularity stems from its characteristics (Wigner, 1932):

- It has a close connection to the marginal probability distributions characterizing the probabilities of the outcomes of von Neumann measurements of the system.

- It lends itself to a visualization of quantum states, and some of their properties.

- It is a versatile calculation tool for normally ordered operators.

With the use of this distribution function, it is straightforward to cast quantum mechanics in a form which resembles the classical theory of statistical averages over the classical phase space, with the Wigner distribution function playing a role analogous to a probability function

---

[1] Wigner's original motivation for introducing it, was to be able to calculate the quantum correction to the second virial coefficient of a gas, which indicates how it deviates from the ideal gas law (Wigner, 1932). Classically, to calculate the second virial coefficient one needs a joint distribution of position and momentum. So Wigner devised the simplest joint distribution that gave, as marginals, the quantum mechanical distributions of position and momentum. The quantum mechanics came in the distribution, but the distribution was used in the classical manner. It was a hybrid method. Also, Wigner was motivated in part by the work of Kirkwood (Kirkwood, 1933) who had previously calculated this quantity but Wigner improved it.

(Fairlie, 1964). Consequently, the Wigner distribution function has been used extensively to study the classical limit of quantum mechanical systems (Kim & Noz, 1991; Mayer & Band, 1947; Moyal, 1949).

There is in principle an infinite variety of quantum phase-space distribution functions corresponding to an infinite number of possible ordering rules of two noncommuting operators and their linear combinations (Lee, 1995). The general class of this distributions is given by (Cahill & Glauber, 1969; Cohen, 1966; Kakazu et al., 2007). Distribution functions in general have different properties and are associated with various dynamical equations, so they may be described most conveniently by distribution functions having different characteristics (Hillery et al., 1984). Other distribution functions that have been considered in the past include those of Kirkwood (Kirkwood, 1933), Margenau-Hill (Johansen & Luis, 2004; Margenau & Hill, 1961; Terletsky, 1937)[2], Husimi (Husimi, 1940), Q-functions (Husimi, 1940; Kano, 1965; Smith, 2006), Page (Page, 1952), Glauber-Sudarshan (Glauber, 1963; Sudarshan, 1963), Rihaczek (Rihaczek, 1968), and Choi-Williams (Choi & Williams, 1989). Another very used function is the ambiguity function (Bastiaans, 1980; Marks & Hall, 1979; Woodward, 1963).

Phase space representation, in particular through the so-called *Wigner Distribution Function*, has proven to be a very effective tool applied in many branches of physics (Cohen, 1995; Kim & Wigner, 1987; Kim & Noz, 1991; Mecklenbrauker et al., 1997; Moyal, 1949; Stewart et al., 2002; Wigner, 1932), and more specifically in fiber optics (Dragoman & Meunier, 1998; Kominis & Hizanidis, 2002; Reyes et al., 1999a; Rivera & Castano, 2010a; Sheppard & Larkin, 2000; Voss et al., 1999). The Wigner distribution function was invented by Wigner (Wigner, 1932) to study the quantum corrections to the classical behavior of certain statistical systems described by the Boltzmann formula. For the evaluation of the Wigner function are various implementations (Bala & Prabhu, 1989; Easton et al., 84; Eilouti & Khadra, 1989; Flandrin et al., 1984; Frank et al., 2000; Gupta & Asakura, 1986; Lohmann, 1980; Lopez et al., 2002; Maanen, 1985; Mateeva & Sharlandjiev, 1986; Rivera et al., 1997; Subotic & Saleh, 1984).

In this chapter the notation will be for the optical position coordinates $\mathbf{q}$ (that corresponds to the interaction of the ray with a $z = 0$ reference plane), and for the canonically conjugate momentum $\mathbf{p}$ (which describes the direction of the ray with respect to the normal at the point $\mathbf{q}$ that evolves over the system's optical axis, z) (Buchdahl, 1970).

Consider a particle in one dimension. Classically, the particle is described by a *phase space distribution* $P_{cl}(\mathbf{q}, \mathbf{p})$. The average of a function of position and momentum $A(\mathbf{q}, \mathbf{p})$ can then be expressed as

$$\langle A \rangle_{cl} = \int_{-\infty}^{\infty} d\mathbf{q} \int_{-\infty}^{\infty} d\mathbf{p}\, A(\mathbf{q}, \mathbf{p})\, P_{cl}(\mathbf{q}, \mathbf{p}). \tag{1}$$

A quantum mechanical particle is described by a density matrix $\hat{\rho}$, and the average of a function of the position and momentum operators $\hat{A}(\hat{q}, \hat{p})$ as

$$\langle A \rangle_{quant} = \mathrm{Tr}\,(\hat{A}\,\hat{\rho}). \tag{2}$$

It must be admitted that, given the classical expression $A(\mathbf{q}, \mathbf{p})$, the corresponding self adjoin operator $\hat{A}$ is not uniquely defined. The use of a *quasiprobability phase space distribution*

---

[2] Kirkwood attempted to extend the classical theory to the quantum case and devised the distribution commonly called the Rihaczek or Margenau-Hill distribution to do that. Many years later, Margenau and Hill derived the distribution that bears their name. The importance of the Margenau-Hill work is not the distribution but the derivation. They were also the first to consider joint distributions involving spin.

$P_Q(\mathbf{q}, \mathbf{p})$, however, does give such a definition by expressing the quantum mechanical average as

$$\langle A \rangle_{quant} = \int_{-\infty}^{\infty} d\mathbf{q} \int_{-\infty}^{\infty} d\mathbf{p} \, A(\mathbf{q}, \mathbf{p}) \, P_Q(\mathbf{q}, \mathbf{p}) , \tag{3}$$

where the function $A(\mathbf{q}, \mathbf{p})$ can be derived from the operator $\hat{A}(\hat{q}, \hat{p})$ by a well defined correspondence rule. This allows one to cast quantum mechanical results into a form in which they resemble classical ones. This is a reformulation of Schrödinger quantum mechanics which describes states by functions in configuration space (Kim & Noz, 1991).

In the case where P_Q in (3) is chosen to be the Wigner Distribution function (Wigner, 1932), then the correspondence between $A(\mathbf{q}, \mathbf{p})$ and $\hat{A}$ is that proposed by Weyl (Weyl, 1927), as was first demonstrated by Moyal (Moyal, 1949).

The requirement given by Eq. (3) let us to define a function in the $6N$ dimensional $\mathbf{q}, \mathbf{p}$ phase space, called the *Wigner Distribution Function* in terms of the density matrix, ρ as:

$$\mathcal{W}_\rho(\mathbf{q}, \mathbf{p}; t) \equiv \left(\frac{1}{\pi\hbar}\right)^{3N} \int_{-\infty}^{\infty} d\mathbf{r} \, \exp\left(\frac{2i}{\hbar} \mathbf{p} \cdot \mathbf{r}\right) \rho(\mathbf{q} - \mathbf{r}, \mathbf{q} + \mathbf{r}; t) , \tag{4}$$

Because for pure states described by a wavefunction Ψ, the density matrix is given by (vonNeumann, 1927)

$$\rho(\mathbf{q}, \mathbf{q}') = \Psi^*(\mathbf{q}') \Psi(\mathbf{q}) , \tag{5}$$

the expression (4) for pure states can be rewritten in coordinate representation as:

$$\mathcal{W}_\Psi(\mathbf{q}, \mathbf{p}; z) \equiv \frac{1}{2\pi\hbar} \int_{-\infty}^{\infty} d\mathbf{r} \, \Psi^*\left(\mathbf{q} - \frac{1}{2}\mathbf{r}; z\right) e^{-i[\mathbf{p} \cdot \mathbf{r}]/\hbar} \Psi\left(\mathbf{q} + \frac{1}{2}f\mathbf{r}; z\right) \tag{6}$$

or taking the Fourier transform (Goodman, 1968) in momentum representation as

$$\mathcal{W}_\Psi(\mathbf{q}, \mathbf{p}; z) = \frac{1}{2\pi\hbar} \int_{-\infty}^{\infty} d\mathbf{r} \, \tilde{\Psi}\left(\mathbf{p} + \frac{1}{2}\mathbf{r}; z\right) e^{-i[\mathbf{q} \cdot \mathbf{r}]/\hbar} \tilde{\Psi}^*\left(\mathbf{p} - \frac{1}{2}\mathbf{r}; z\right) , \tag{7}$$

where $\hbar$ is the Planck constant divided by 2π, $\tilde{\Psi}$ denotes the Fourier transform of Ψ and the asterisk represents the complex conjugate. In geometric optics, $\hbar$ corresponds to the wavelength λ of the beam (Wolf, 2004).

In Wigner phase-space representation everything we have said for the coordinate domain holds for the momentum domain because the Wigner distribution is basically identical in form in both domains (compare equations 6 and 7). The complete symmetry between $\mathbf{q}$ and $\mathbf{p}$ in the former definitions of the Wigner function (equations 6 and 7), indicates that space and momentum have equal weight in this description (Moyal, 1949). Due to this, the Wigner distribution function can be thought as the expected value of the parity operator around $(\mathbf{q}, \mathbf{p})$ in the phase space (Royer, 1997); i.e. the Wigner function is proportional to the overlap of $\Psi(\mathbf{q}, z)$ with its specular image around $(\mathbf{q}, \mathbf{p})$, that is a measure of "how much centered" is $\Psi(\mathbf{q}, z)$. Note that the Wigner distribution function is a 4-dimensional phase space distribution function, where two dimensions correspond to real space and the other two to momentum space. The Wigner distribution function is a real function that can take either positive and negative values, however, only for a Gaussian the Wigner distribution function is positive everywhere (Hudson, 1974; Soto & Claverie, 1983); therefore, one cannot interpret it as a classical probability function in phase space (Lee, 1995). Tthe value of $\mathcal{W}_\Psi$ mirrors closely the intuitive objects in the model, that in the case of quantum optics may be the coherent states of the radiation field (Glauber, 1965), and in monochromatic paraxial wave optics, they are often beams with Gaussian position and inclination distributions (Hillery et al., 1984; Rivera & Castano, 2010).

From all phase space representations, the Wigner Distribution Function can be uniquely distinguished (among shift-invariant joint distributions) by imposing a requirement of correct marginals with respect to arbitrary directions in the time-frequency plane, thus connecting the Wigner distribution with the fractional Fourier transform (Atakishiyev et al., 1999). It also contains all the information of the system and it can be proved that it contains the hologram of the signal (Wolf & Rivera, 1997).

For numerical calculations it is very useful to note that the Wigner distribution function is the Fourier transform of the kernel (Wigner, 1932):

$$\mathcal{W}_{\Psi}(\mathbf{q}, \mathbf{p}; z) = \Psi\left(\mathbf{q} + \frac{1}{2}\mathbf{r}\right) \Psi^*\left(\mathbf{q} - \frac{1}{2}\mathbf{r}\right). \tag{8}$$

Because $\mathcal{W}_{\Psi}(\mathbf{q}, \mathbf{p}; z)$ is Hermitian $[\mathcal{W}_{\Psi}(\mathbf{q}, \mathbf{r}) = \mathcal{W}_{\Psi}^*(\mathbf{q}, -\mathbf{r})]$, the Wigner distribution function is real (Moyal, 1949).

When we integrate $\mathcal{W}_{\Psi}(\mathbf{q}, \mathbf{p})$ over $\mathbf{p}$, we obtain the probability distribution in $\mathbf{q}$, while if we integrate $\mathcal{W}_{\Psi}(\mathbf{q}, \mathbf{p})$ over $\mathbf{q}$, we obtain the probability distribution in $\mathbf{p}$, (Moyal, 1949). Then, to recover either the image $|\Psi(\mathbf{q}; z)|^2$ (light intensity on the two-dimensional screen of coordinate $\mathbf{q}$ at the optical axis position z) or the diffraction pattern $|\tilde{\Psi}(\mathbf{p}; z)|^2$, it is necessary to make a simple projection of the Wigner distribution function (Wigner, 1932):

$$|\Psi(\mathbf{q}; z)|^2 = \int_{-\infty}^{\infty} d\mathbf{p}\, \mathcal{W}_{\Psi}(\mathbf{q}, \mathbf{p}; z), \tag{9}$$

$$|\tilde{\Psi}(\mathbf{p}; z)|^2 = \int_{-\infty}^{\infty} d\mathbf{q}\, \mathcal{W}_{\Psi}(\mathbf{q}, \mathbf{p}; z). \tag{10}$$

If the signal or image of interest is nonstationary, the Wigner distribution function gives the local spectrum centered at $\mathbf{p}$ as a function of location (Bartelt et al., 1980). Thus, the total energy of $\Psi(\mathbf{q}, z)$ can be obtained from integration of $\mathcal{W}_{\Psi}(\mathbf{q}, \mathbf{p}; z)$ over the entire phase space (Hillery et al., 1984).

Moreover $|\mathcal{W}_{\Psi}(\mathbf{q}, \mathbf{p}; z)| \leq (2\pi\hbar)^{-1}$.

Another interesting property (Schempp, 1986) is that the Wigner distribution function has the same extension and is band-limited as the function $\Psi(\mathbf{q}, z)$.

The Wigner distribution function is the expectation value of the parity operator about the phase-space point $\mathbf{q}, \mathbf{p}$ (Royer, 1997). To show this, let us first rewrite

$$\mathcal{W}(\mathbf{q}, \mathbf{p}) = \left(\frac{1}{\pi\hbar}\right)^{3N} \langle\Psi|\hat{\Pi}_{\mathbf{q},\mathbf{p}}|\Psi\rangle, \tag{11}$$

where the operator $\hat{\Pi}_{\mathbf{q},\mathbf{p}}$ has the following three equivalent expressions:

$$\hat{\Pi}_{\mathbf{q},\mathbf{p}} = \int_{-\infty}^{\infty} dr\, e^{2ipr/\hbar}\, |q - r\rangle\, \langle q + r|,$$

$$= \int_{-\infty}^{\infty} dk\, e^{-2ikq/\hbar}\, |p + k\rangle\, \langle p - k|,$$

$$= \left(\frac{1}{\pi\hbar}\right)^{3N} \int_{-\infty}^{\infty} dk \int_{-\infty}^{\infty} dr\, e^{i[k(\hat{R}-q)+r(\hat{P}-p)]/\hbar}. \tag{12}$$

Let us now consider the special case $q = 0, p = 0$, and denote $\hat{\Pi}_{q=0,p=0} = \hat{\Pi}$; we have

$$\hat{\Pi} = \int_{-\infty}^{\infty} dq \, |-q\rangle \, \langle q| \, ,$$

$$= \int_{-\infty}^{\infty} dp \, |p\rangle \, \langle -p| \, ,$$

$$= \left(\frac{1}{\pi\hbar}\right)^{3N} \int_{-\infty}^{\infty} dk \int_{-\infty}^{\infty} dy \, e^{i[k\hat{R}+y\hat{P}]/\hbar} \, . \tag{13}$$

From (13) it is immediately apparent that $\hat{\Pi}$ is the parity operator (about the origin): it changes $\Psi(\mathbf{q})$ into $\Psi(-\mathbf{q})$ and $\tilde{\Psi}(\mathbf{p})$ into $\tilde{\Psi}(-\mathbf{p})$, or equivalently

$$\hat{\Pi}\hat{R}\hat{\Pi} = -\hat{R} \, , \qquad \hat{\Pi}\hat{P}\hat{\Pi} = -\hat{P} \, , \tag{14}$$

moreover,

$$\hat{\Pi}^{-1} = \hat{\Pi} \, . \tag{15}$$

We now observe that $\hat{\Pi}_{\mathbf{q},\mathbf{p}}$ may be obtained from $\hat{\Pi}$ by a unitary transformation

$$\hat{\Pi}_{\mathbf{q},\mathbf{p}} = \hat{D}(\mathbf{q},\mathbf{p}) \, \hat{\Pi} \, \hat{D}(\mathbf{q},\mathbf{p})^{-1} \, ; \tag{16}$$

here

$$\hat{D}(\mathbf{q},\mathbf{p}) = e^{i(p\hat{R}-q\hat{P})/\hbar} \tag{17}$$

is a phase-space displacement operator, introduced by Glauber (Glauber, 1963) in connection with a different, though related, type of phase-space representation of quantum mechanics, the coherent-state representation. We have the actions

$$\hat{D}(\mathbf{q},\mathbf{p})^{-1} \, \hat{R} \, \hat{D}(\mathbf{q},\mathbf{p}) = \hat{R} + q \, , \tag{18}$$

$$\hat{D}(\mathbf{q},\mathbf{p})^{-1} \, \hat{P} \, \hat{D}(\mathbf{q},\mathbf{p}) = \hat{P} + p \, , \tag{19}$$

$$\hat{D}(\mathbf{q},\mathbf{p})^{-1} \, F(\hat{R},\hat{P}) \, \hat{D}(\mathbf{q},\mathbf{p}) = F(\hat{R}+q, \hat{P}+p) \, . \tag{20}$$

From this follows directly

$$\hat{\Pi}_{\mathbf{q},\mathbf{p}}(\hat{R} - q)\hat{\Pi}_{\mathbf{q},\mathbf{p}} = -(\hat{R} - q) \, , \tag{21}$$

$$\hat{\Pi}_{\mathbf{q},\mathbf{p}}(\hat{P} - p)\hat{\Pi}_{\mathbf{q},\mathbf{p}} = -(\hat{P} - p) \, , \tag{22}$$

that is, $\hat{\Pi}_{\mathbf{q},\mathbf{p}}$ reflects about the phase-space point $\mathbf{q}, \mathbf{p}$ and is thus the parity operator about that point. Note that

$$(\hat{\Pi}_{\mathbf{q},\mathbf{p}})^2 = 1 \, . \tag{23}$$

The Wigner function, is thus $\left(\frac{1}{\pi\hbar}\right)^{3N}$ times the expectation value of the parity operator about $\mathbf{q}, \mathbf{p}$. Alternatively, $\mathcal{W}(\mathbf{q},\mathbf{p})$ is proportional to the overlap of Ψ with its mirror image about $\mathbf{q}, \mathbf{p}$, which is clearly a measure of how much Ψ is "centered" about $\mathbf{q}, \mathbf{p}$.

3. Light propagation on a fiber

Propagation of light in a fiber is governed by Maxwell equations (Born & Wolf, 1999). Consider a monocromatic light beam of frequency ω propagating through a fiber of refractive index n, described by the wavefunction $\Psi(x, y, z)$. It can be shown that this beam obeys the Helmholtz equation (Born & Wolf, 1999):

$$\frac{\partial^2 \Psi}{\partial x^2} + \frac{\partial^2 \Psi}{\partial y^2} + \frac{\partial^2 \Psi}{\partial z^2} + \frac{\omega^2}{c^2} n^2 \Psi = 0 , \tag{24}$$

where $n = n(x, y, z)$ is the refractive index of the fiber.

Under paraxial approach, the beam is almost parallel and close to the optical axis of the system, z, then $n = n(0, 0, z)$, and Ψ vary slowly with z allowing to neglect second order derivatives in the z direction. This considerations let to write equation (24) as the parabolic type equation (Leontovich & Fock, 1946)

$$\frac{i}{k} \frac{\partial \Psi}{\partial \tau} = \frac{1}{2k^2} \left(\frac{\partial^2 \Psi}{\partial x^2} + \frac{\partial^2 \Psi}{\partial y^2} \right) + \left(n_0^2 - n^2 \right) \Psi , \tag{25}$$

where n_0 is the vacuum refractive index, and

$$\tau = - \int_0^z \frac{1}{n_0(z')} dz' . \tag{26}$$

Equation (25) shows that the light beam propagation in the paraxial approximation is described by a Schrödinger equation where the wavelenght $\lambda = \frac{1}{k}$ plays the role of the Planck constant and instead of time appears z. The potential well is given by the refractive index $n_0^2 - n^2(x, y, z)$. This treatment translate the problem of solving the Helmholtz equation (24) to solve the Schrödinger equation for a system with two degrees of freedom (x, y) in a time-dependent (z) potential well.

This Schrödinger equation (25) is valid for any wave that follows the Helmholtz equation under the paraxial approach (for a detailed description check (Arnaud, 1976; Manko, 1986; Marcuse, 1972). The validity of this approximation can be verified using the moments of the Wigner distribution function of the solution as shown in (Rivera et al., 1997). To solve the problem in fiber optics it can be applied the formalism of symplectic groups through coherent state representation of quantum mechanics (Manko & Wolf, 1985).

In general, the output Wigner function of an optical system is related to the input Wigner through (Castano et al., 1982; Gutierrez & Castano, 1992):

$$\mathcal{W}_{\Psi_{out}}(\mathbf{q}, \mathbf{p}; z) = \mathcal{W}_{\Psi_{in}}(a \mathbf{q} + b \mathbf{p}, c \mathbf{q} + d \mathbf{p}; z) , \tag{27}$$

where a, b, c and d are parameters which depend on the specific system under study.

As an example, the free space propagator is given by

$$\mathcal{W}_{\Psi_{out}}(\mathbf{q}, \mathbf{p}; z) = \mathcal{W}_{\Psi_{in}} \left(\mathbf{q} - \frac{z}{k} \mathbf{p}, \mathbf{p}; z \right) , \tag{28}$$

for a lens of focal length f, we have

$$\mathcal{W}_{\Psi_{out}}(\mathbf{q}, \mathbf{p}; z) = \mathcal{W}_{\Psi_{in}} \left(\mathbf{q}, \frac{1}{f} \mathbf{q} + \mathbf{p}; z \right) , \tag{29}$$

and to obtain a Fourier transform we use:

$$\mathcal{W}_{\Psi_{out}}(\mathbf{q}, \mathbf{p}; z) = \mathcal{W}_{\Psi_{in}} (-\mathbf{p}, \mathbf{q}; z) . \tag{30}$$

4. Gaussian beam propagation in optical fibers

To model optical fibers it is common to consider gaussian beams that travel freely through space (Rivera & Castano, 2010). Gaussians are also ubiquitous in quantum mechanics, where they are intimately related to the harmonic oscillator (Gitterman, 2003; Moshinsky, 1996; Sako & Diercksen, 2003), to the coherent (Gori et al., 2003; Grewal, 2002; Grosshans et al., 2003; Lauterborn et al., 1993; Lesurf et al., 1993) and squeezed states formalism (Agarwal & Ponomarenko, 2003; Dodonov, 2002; Kim et al., 2002; Sohma & Hirota, 2003). In Quantum Optics, Gaussian beams are fundamental to test and to compare wave optical models and systems (Berry, 1994; Oraevsky, 1998; Rivera et al., 1997).

Using Fermat minimal action principle, it can be proved that the system is governed by the optical Hamiltonian (Rivera et al., 1995):

$$H = -\sqrt{n^2 - p^2} \,. \tag{31}$$

This Hamiltonian generates a ray path, i.e. a unidimensional group of canonical transformations of the points of the optical phase space. In a three-dimensional optical medium we denote the two screen coordinates (perpendicular to the optical axis) by $\mathbf{q} = (x.y)$ and the optical axis coordinate as z.

When the canonical transformation has a nonlinear part, it is possible to identify this nonlinearity as the effect of aberrations as is studied in (Rivera et al., 1997; Rivera & Castano, 2010). An alternative approach (called coherent states for Lie groups) uses the continuous representations in quantum mechanics as a particular case of arbitrary Lie groups and can be used in fiber optics for analyzing nonquadratic media under the action of Hamiltonians that are the linear form of the Lie group representation with z dependent coefficients (Klauder, 1964). In geometric optics (paraxial approach), momentum is $|\mathbf{p}| = n \sin \theta$, where n denotes the refractive index and θ is the angle between the ray and the optical axis (Wolf, 2004).

A Gaussian function Γ associated to the one-dimensional real coordinate x is defined as (Simon, 2002)

$$\Gamma(x) = M \exp \left[-\frac{(x - x_0)^2}{2w_0} + ip_0 x \right] , \tag{32}$$

where $M = \left(\frac{w_1}{\pi |w_0|^2} \right)^{1/4}$, x_0, p_0 are real numbers, and $w_0 = w_1 + iw_2$, $w_1 > 0$ is a complex number. The dimension of w_0 is $[x^2]$, the one of x_0 is $[x]$, and that of p_0 is $[x^{-1}]$. The pre-exponential factor M guarantees the normalization condition

$$\langle \Gamma \mid \Gamma \rangle = \int_{-\infty}^{\infty} dx \, \Gamma^*(x) \Gamma(x) = 1 \,. \tag{33}$$

This Gaussian is centered at x_0 and has a complex width $\sqrt{2w_0}$. The value at its maximum is M. If $p_0 \neq 0$ or $w_2 \neq 0$, this Gaussian shows oscillations.

The Fourier transform of the Gaussian Γ (Equation 32) provides the momentum representation of the beam (Goodman, 1968):

$$\tilde{\Gamma}(p) = \frac{1}{\sqrt{2\pi}} \int_{-\infty}^{\infty} dx \, e^{-ipx} \Gamma(x) = \left(\frac{w_1}{\pi} \right)^{\frac{1}{4}} \exp \left[-\frac{w_0(p - p_0)^2}{2} - ix_0(p - p_0) \right] . \tag{34}$$

Interestingly, it is another Gaussian, centered in p_0, with width $\sqrt{2/w_0}$ and it oscillates for $x_0 \neq 0$.

In phase space, the Gaussian Γ, is represented by its Wigner distribution function (Rivera & Castano, 2010):

$$\mathcal{W}_\Gamma(x,p) = 2\exp\left\{-\frac{(x-x_0)^2}{w_1} - \frac{|w_0|^2}{w_1}(p-p_0)^2 + \frac{2w_2}{w_1}(x-x_0)(p-p_0)\right\}, \qquad (35)$$

that is a two-dimensional Gaussian; coordinate centered at x_0 with width $\sqrt{w_1}$, momentum center p_0 with width $|w_0|/\sqrt{w_1}$, and tilted by $\arctan(2w_2/w_1)$.

A *Vacuum Coherent State* (Dodonov, 2002) is a Gaussian function with $x_0 = p_0 = 0$ and $w_0 = 1$. It has the important property of being the only state described by the same function in both coordinate and momentum representation (Dodonov & Manko, 2000). A generalized *coherent state* is described by a Gaussian function with $w_0 = 1$, but x_0 and p_0 arbitrary. The state for which $w_0 \neq 1$ is called a *squeezed state* (Dodonov, 2002).

To evaluate the Wigner distribution function of a sectioned fiber we assume that an optical fiber is a cylinder of radius a and infinite length. We will consider a Gaussian traveling through a fiber with constant refractive index n until a break generated by a section with different refractive index m encapsulated by two parabolic surfaces. The Symplectic Map for this system is given by the product of the initial propagator (before the break), first refraction, propagation between refraction surface, second refraction (after the break), and last propagation. It can be shown that the Symplectic Map of this system is (Reyes & Castano, 2000)

$$M_{total} = \dots e^{:F_4:}\, e^{:F_2:}, \qquad (36)$$

where

$$e^{:F_2:} = e^{-\frac{a}{2n}\sin(2k\pi z):\vec{\mathbf{p}}^2:}e^{\alpha(n-m):\vec{\mathbf{q}}^2:}e^{-\frac{\gamma}{2m}:\vec{\mathbf{p}}^2:}$$
$$\times e^{-\frac{\gamma}{2}:\vec{\mathbf{p}}^2:}e^{-\alpha(n-m):\vec{\mathbf{q}}^2:}e^{-\frac{a}{2n}\sin(2k\pi z'):\vec{\mathbf{p}}^2:},$$

$$e^{:F_4:} = e^{\mathcal{A}:(\vec{\mathbf{p}}^2)^2:+\mathcal{B}:\vec{\mathbf{p}}^2(\vec{\mathbf{p}}\cdot\vec{\mathbf{q}}):+\mathcal{C}:(\vec{\mathbf{p}}\cdot\vec{\mathbf{q}})^2:+\mathcal{D}:\vec{\mathbf{p}}^2\vec{\mathbf{q}}^2:}$$
$$\times e^{\mathcal{E}:\vec{\mathbf{q}}^2(\vec{\mathbf{p}}\cdot\vec{\mathbf{q}}):+\mathcal{F}:(\vec{\mathbf{q}}^2)^2:}.$$

Here, z and z' give the propagation before and after the break, respectively.

In Eq. (36), the exponential $e^{:F_2:}$ is the Gaussian term, while $e^{:F_4:}$ corresponds to the aberration term. This method simplifies the optical problem of obtaining the image of an optical system to the determination of the corresponding Symplectic Map associated to the system, thus reducing the problem to simple matrices products.

In order to calculate the Wigner distribution function we need to use the following correspondence

$$:\vec{\mathbf{p}}: \longrightarrow \vec{\mathbf{p}}$$
$$:\vec{\mathbf{q}}: \longrightarrow \vec{\mathbf{q}}$$

to make the Symplectic map, and consider the convolution between a point source and M_{total}. The point source is defined by the Dirac Delta

$$F(\vec{\mathbf{p}'},\vec{\mathbf{q}'}) = \delta(\vec{\mathbf{q}'}-\vec{\mathbf{q}},\vec{\mathbf{p}'}-\vec{\mathbf{p}}). \qquad (37)$$

The convolution between F and M_{total} (up to fourth order) is

$$F * M_{total} = F(\vec{\mathbf{p}'},\vec{\mathbf{q}'}) * e^{F_2}e^{F_4}\dots$$
$$\simeq [F(\vec{\mathbf{p}'},\vec{\mathbf{q}'}) * e^{F_2}](1+F_4)F(\vec{\mathbf{p}'},\vec{\mathbf{q}'}) * e^{F_2}$$
$$+[F(\vec{\mathbf{p}'},\vec{\mathbf{q}'}) * e^{F_2}]F_4. \qquad (38)$$

The aberration of the system respect to the Gaussian ray (second term of the last equation) is

$$F * e^{F_2} = F(\vec{p}', \vec{q}') * e^{F_2} = C_1 \frac{\pi}{\sqrt{\alpha_5}} e^{-\left(\frac{\alpha_1 \vec{p}'^2 - \alpha_2 \vec{p}' \cdot \vec{p} + \alpha_3 \vec{p}^2}{\alpha_4}\right)},$$

where

$$\alpha_1 = \frac{a}{2n} \sin(2k\pi z')(\beta + \frac{\gamma}{2}),$$

$$\alpha_2 = \beta \frac{a}{2n} \sin(2k\pi z'),$$

$$\alpha_3 = \beta(\frac{\gamma}{2} + \frac{a}{2n} \sin(2k\pi z')),$$

$$\alpha_4 = \beta + \frac{\gamma}{2} + \frac{a}{2n} \sin(2k\pi z'),$$

$$\alpha_5 = (\frac{\gamma}{2n} + \frac{a}{2n} \sin(2k\pi z))(\beta + \frac{\gamma}{2n} + \frac{a}{2n} \sin(2k\pi z')).$$

Without perturbation ($\gamma \to 0$), the aberration yield

$$F * e^{F_2} = C_1 \frac{\pi}{\sqrt{\frac{a}{2n} \sin(2k\pi z)}},$$

that corresponds to the convolution between $e^{-\frac{a}{2n} \sin(2k\pi z)\vec{p}^2}$ and a point source.

Now we can calculate the Wigner distribution function in the image plane substituting in Equation (6) the Symplectic Map of this system, Eqs. (36) and (38):

$$W(\vec{q}', \vec{p}') = \int_{-\infty}^{\infty} d\vec{r}\, e^{-i\vec{p}' \cdot \vec{r}}$$

$$\times \left\{ F * e^{F_2}(\vec{q}' + \frac{\vec{r}}{2}) + [F * e^{F_2}]F_4(\vec{q}' + \frac{\vec{r}}{2}) \right\}$$

$$\times \left\{ F * e^{F_2}(\vec{q}' - \frac{\vec{r}}{2}) + [F * e^{F_2}]F_4(\vec{q}' - \frac{\vec{r}}{2}) \right\},$$

that can be rewritten up to fourth order as

$$W(\vec{q}', \vec{p}') =$$

$$C_1^2 \frac{2\pi^3}{\alpha_5} e^{i\pi} e^{-(2/\alpha_4)(\alpha_1 \vec{p}''^2 - [2\alpha_1 + \alpha_2]\vec{p}'' \cdot \vec{p} + [\alpha_1 + \alpha_2 + \alpha_3]\vec{p}^2)}$$

$$\times \left\{ 1 + 2F_4\delta(\vec{p}') + \pi^2 F \left(\frac{d^2}{dp_1'^2} + \frac{d^2}{dp_2'^2} \right)^2 + 8\pi^2 e^{i\pi} \right.$$

$$\times \left[C \left(p_1' \frac{d}{dp_1'} + p_2' \frac{d}{dp_2'} \right)^2 + 2F \left(q_1' \frac{d}{dp_1'} + q_2' \frac{d}{dp_2'} \right)^2 \right.$$

$$+\mathcal{E} \left(p_1' q_1' \frac{d^2}{dp_1'^2} + p_2' q_2' \frac{d^2}{dp_2'^2} + (p_2' q_1' + p_1' q_2') \frac{d}{dp_1'} \frac{d}{dp_2'} \right)$$

$$\left. \left. + \frac{1}{2}[\mathcal{D}(\vec{p}') + \mathcal{E}(\vec{p}') \cdot \vec{q}'] \left(\frac{d^2}{dp_1'^2} + \frac{d^2}{dp_2'^2} \right) \right] \right\} \delta(\vec{p}'),$$

$$(39)$$

with $\vec{\mathbf{p}}' = \vec{\mathbf{p}}'' - \vec{\mathbf{p}}$. From this equation is clear that after the break, the Wigner distribution function is a Gaussian with center in $(2\alpha_1 + \alpha_2\vec{\mathbf{p}})/(2\alpha_1)$. The Wigner function found have a general phase of 2π, except in the term $(4\pi^2/3)\mathcal{F}\{\frac{d^2}{dp_1'^2} + \frac{d^2}{dp_2'^2}\}^2\delta(\vec{\mathbf{p}}')$ where the phase is π. Thus the initial Gaussian is modified by a corrective term, the F_4 polynomial (the exponent of the Aberration Lie Operator).

The limiting case without break, $\gamma \to 0$, has a Wigner distribution function given by

$$W(\vec{\mathbf{q}}', \vec{\mathbf{p}}') = \frac{C_1^2 e^{i\pi}}{\frac{a}{2n}sin(2k\pi z)}\delta(\vec{\mathbf{p}}')\,(1 + 2F_4 + \ldots) \tag{40}$$

5. Conclusions

As shown in this chapter, phase space approach (through the Wigner distribution function) simplifies the calculation and helps in the description of optical fibers. This is due to the fact that in phase space representation, the relevant properties of the system can be obtained by simple matrices products. Quasiprobability distribution functions are useful not only as calculation tools but can also provide insights into the connections between geometric opticsl and wave optics due to the fact that they allow one to express wave optics averages in a form which is very similar to that for classical averages. In this sense it serves to validate paraxial approximation.

6. References

Agarwal, G. S. & Ponomarenko, S.A. (2003). Minimum-correlation mixed quantum states. *Physical Review A*, Vol. 67, article 032103, ISSN: 1050-2947.

Allen, R. L. & Mills, D. (2004). *Signal Analysis: Time, Frequency, Scale, and Structure*. Wiley, Berlin.

Arnaud J. A. (1976). *Beam and Fiber Optics*, Academic Press, New York.

Atakishiyev, N.; Vicent, L. E. & Wolf, K. B. (1999). Continuous vs. discrete fractional Fourier transforms. *J. Comput. Appl. Math*. Vol. 107, pp. 73.

Bakhturin, Y. (2003). *Groups, Rings, Lie and Hopf Algebras*. Kluwer Academic Publishers, New York.

Bala, S. & Prabhu, K. M. M. (1989). New method of computing Wigner-Ville distribution. *Electr. Letter*, Vol. 25, pp. 336.

Bao, X. & Chen, L. (2011). Recent Progress in Brillouin Scattering Based Fiber Sensors *Sensors*, Vol. 11, pp. 4152-4187.

Bastiaans, M. J. (1980). Wigner distribution function display: a supplement to ambiguity function display using a 1-D input. *Applied Optics*. Vol. 19, No. 2, pp. 192-193, ISSN: 1559-128X.

Balazs, N. L. & Jennings, B. K. (1984) Wigner's function and other distribution functions in Mock phase space. *Physics Reports*, Vol. 104, No. 6, pp. 347-391, ISSN: 0370-1573.

Bartelt, H.O.; Brenner, K.-H. & Lohmann, A. H. (1980). The Wigner distribution function and its optical production. *Optics Communications*, Vol. 32, pp. 32-38, ISSN: 0030-4018.

Benabid, F. & Roberts, P. J. (2011). Linear and nonlinear optical properties of hollow core photonic crystal fiber. *Journal of Modern Optics* Vol. 58, No. 2, pp. 87-124.

Benchekh, K.; Gravier, F.; Douady, J.; Levenson, A. & Boulanger, B. (2007). Triple photons: a challenge in nonlinear and quantum optics. *C. R. Physique*, Vol. 8, pp. 206âĂŞ220.

Berry, M. V. (1977). Semiclassical Mechanics in Phase Space: A Study of Wigner's Function. *Philosophical Transations of the Royal Society of London A* Vol. 287, No. 1343, pp. 237-271.

Berry, M. V. (1994). Evanescent and real waves in quantum billiards and Gaussian beams *Journal of Physics A: Mathematical and Theoretical* Vol. 27, No. 11, pp. L391-L398, ISSN 1751-8113.

Boashash, B. (2003). *Time Frequency Analysis*. Elsevier Science, Holland.

Born, M. & Wolf, E. (1999). *Principles of Optics*, 7th ed. Cambridge University Press, ISBN 0 521 642221, Cambridge.

Bottacchi, M. (2002). *Optical Fibre Transmission Theory, Technology and Design: Optic Propagation Theory*. John Wiley & Sons, New York.

Breitenbach, G.; Muller, T.; Pereira, S. F.; Poizat, J. P.; Schiller, S. & Mlynek, J. (1995). Squeezed vacuum from a monolithic optical parametric oscillator, *Journal of the Optical Society of America B*, Vol.12, pp. 2304.

Breitenbach, G.; Schiller, S. & Mlynek J. (1997). Measurement of the quantum states of squeezed light. *Nature* Vol. 387, pp. 471, ISSN: 0028-0836.

Buchdahl, H. A. (1970). *An introduction to Hamiltonian Optics*. Cambridge University Press, Cambridge.

Cahill, K. E. & Glauber, R. J. (1969). Density operators and Quasiprobability Distributions. *Physical Review*, Vol. 177, pp. 1882-1896.

Castano, V. M.; Vázquez-Polo, G. & Gutiérrez Castrejón, R. (1982). *Scanning Microscopy Supplement* Vol. 6, pp. 415-418, ISSN: 0892-953X.

Choi, H. I. & Williams, W. J. (1989). Improved time-frequency representation of multicomponent signals using exponential kernels. IEEE Trans. on Acoustics, Speech, and Signal Processing, Vol. 37, pp. 862.

Cohen, L. (1966). Generalized phase-space distribution functions. *Journal of Mathematical Physics*. Vol. 7, pp. 781.

Cohen, L. (1995). *Time-frequency analysis*. Prentice Hall, New Jersey.

Corney, J. F.; Heersink, J.; Dong, R.; Josse, V.; Drummond, P. D.; Leuchs, G. & Andersen, U. L. (2008). Simulations and experiments on polarization squeezing in optical fiber. *Physical Review A*, Vol. 78, article 023831, ISSN: 1050-2947.

Culshaw, B. (1997). *Optical Fiber Sensors: Applications, Analysis and Future Trends*. Artech House, New York.

Dirac, P. A. M. (1935). *The Principles of Quantum Mechanics*. 2nd Ed., Cambridge University Press, Cambridge.

Dragoman, D. (1997). The Wigner Distribution Function in Optics and Optoelectronics. In *Progress in Optics XXXVII*, ed. Wolf, E. Elsevier, Amsterdam.

Dragoman, D. & Meunier, J. P. (1998). Recovery of longitudinally variant refractive index profile from the measurement of the Wigner transform. *Optics Communications*, Vol. 153, pp. 360-367, ISSN: 0030-4018.

Dragt, A. J. & Finn, J. M. (1976). Lie series and invariant functions for analytic symplectic maps. *Journal of Mathematical Physics*, Vol. 17, No. 12, pp. 2215-2227. ISSN: 0022-2488.

Dodonov, V. V. & Manko, O. V. (2000). Universal invariants of quantum-mechanical and optical systems. *Journal of the Optical Society of America A*, Vol.17, No. 12, pp. 2403-2410, ISSN: 1084-7529.

Dodonov, V. V. (2002). Nonclassical states in quantum optics: a squeezed review of the first 75 years, *Journal of Optics B: Quantum and Semiclassical Optics* Vol. 4, No.1, pp. R1, ISSN 1464-4266.

Easton, R. L.; Ticknor, A. J. & Barrett, H. H. (1984). Application of the Radon transform to optical production of the Wigner distribution function. *Optical Engineering.* Vol. 23, pp. 738.

Eilouti, H. H. & Khadra, L. M. (1989). Optimised implementation of real-time discrete Wigner distribution *Electr. Letters,* Vol. 25, pp. 706.

Fairlie, D. B. (1964). The formulation of quantum mechanics in terms of phase space functions. *Proceedings of the Cambridge Philosophical Society,* Vol. 60, No. 3, pp. 581-590, ISSN: 0305-0041.

Filinov, V. S.; Bonitz, M.; Filinov, A. & Golubnychiy, V. O. (2008). *Wigner function quantum molecular dynamics.* Lecture Notes in Physics, Vol. 739, Springer Verlag, Berlin.

Flandrin, P.; Martin, W. & Zakharia, M. (1984). On a hardware implementation of the Wigner-Ville transform, *Proceedings of the International Conference on Digital Signal Processing* 84, pp. 262.

Frank, A.; Rivera, A. L. & Wolf, K. B. (2000). Wigner function of Morse potential eigenstates. *Physical Review A,* Vol. 61, No. 5, article 054102, ISSN: 1050-2947.

Frank, A. & van Isacker, P. (1994). *Algebraic Methods in Molecular and Nuclear Structure.* Wiley Interscience, New York.

Gitterman, M. (2003). Harmonic oscillator with multiplicative noise: Nonmonotonic dependence on the strength and the rate of dichotomous noise. *Physical Review E,* Vol. 67, article 057103, ISSN: 1063-651X.

Glauber, R. J. (1963). Coherent and Incoherent States of the Radiation Field. *Physical Review,* Vol. 131, No. pp. 2766-2788.

Glauber, R. J. (1965). In *Quantum Optics and Electronics,* edited by DeWitt, C.; Blandin A. & Cohen-Tannoudji, C. Gordon and Breach, New York.

Goodman, J. W. (1968). *Introduction to Fourier Optics.* McGraw-Hill, New York.

Gori, F.; Santarsiero, M.; Simon, R.; Piquero, G.; Borghi, R. & Guattari, G. (2003). Coherent-mode decomposition of partially polarized, partially coherent sources. *Journal of the Optical Society of America A* Vol. 20, No.1, pp. 78-84, ISSN: 1084-7529.

Green, H. S. (1951). The Quantum Mechanics of Assemblies of Interacting Particles. *Journal of Chemical Physics,* Vol. 19, pp. 955-956, ISSN 0021-9606.

Grewal, K. S. (2002). Appearance of classical coherent oscillator states under environment-induced decoherence *Physical Review A* Vol. 65, article 052103, ISSN: 1050-2947.

Grochenig, K. (2000). *Foundations of Time-Frequency Analysis (Applied and Numerical Harmonic Analysis).* Birkhäuser, Boston.

Groot, S. R. & Suttorp, L. G. (1972). *Foundations of Electrodynamics.* North-Holland, Amsterdam.

Grosshans, F.; Van Asschet, G.; Wenger, J.; Brouri, R.; Cerf, N.J. & Grangier, P. (2003). Quantum key distribution using gaussian-modulated coherent states. *Nature* Vol. 421, No. 6920, pp. 238-240, ISSN: 0028-0836.

Gupta, A. K. & Asakura, T. (1986). New optical system for the efficient display of Wigner distribution functions using a single object transparency. *Optics Communications,* Vol. 60, pp. 265-271, ISSN: 0030-4018.

Gutierrez Castrejon, R. & Castano, V. M. (1992). *Optik,* Vol. 91, pp. 24, ISSN: 0030-4026.

Hamermesh, M. (1962). *Group Theory and its application to physical problems.* Addison-Wesley Pub., Massachusetts.

Harmon, J. P. et al. (2001). *Optical Polymers: Fibers and Waveguides.* American Chemical Society, New York.

Heisenberg, W. (1930). *The Physical Principles of the Quantum Theory*. 2nd Ed., Cambridge University Press, Cambridge.

Herrmann, R. (1973). *Optik*, Vol. 37, pp. 91, ISSN: 0030-4026.

Hillery, M.; O'Connell, R. F.; Scully, M.O. & Wigner, E. P. (1984). Distribution Functions in Physics: Fundamentals. *Physics Reports*, Vol. 106, No.3, pp. 121-167, ISSN: 0370-1573.

Hudson, R. L. (1974). When is the wigner quasi-probability density non-negative? *Reports on Mathematical Physics*, Vol. 6, No. 2, pp. 249-252, ISSN 0034-4877.

Husimi, K. (1940). Some formal properties of the density matrix. *Prog. Phys. Math. Soc. Japan*, Vol. 22, pp. 264-266.

Jacobson, N. (1979). *Lie Algebras*. Dover Publications, New York.

Johansen, L. M. & Luis, A. (2004). Nonclassicality in weak measurements. *Physical Review A*, Vol. 70, Article 052115, ISSN: 1050-2947.

Kakazu, K.; Kiyuna, M. & Sakai, E. (2007). Systematic Derivation of the Number-Phase Distribution Functions. *Progress on Theoretical Physics* Vol. 118, pp. 827.

Kano, Y. (1965). A new Phase-Space Distribution Function in the Statistical Theory of the Electromagnetic Field. *Journal of Mathematical Physics*, Vol. 6, No. 2, pp. 1913-1921. ISSN: 0022-2488.

Keyl, M. (2002). Fundamentals of quantum information theory. *Physics Reports* Vol. 369, No. 5, pp. 431-548, ISSN: 0370-1573.

Kim, Y. S.& Wigner, E. P. (1987). Covariant phase-space representation for localized light waves. *Physical Review A*, Vol. 36, pp. 1293-1296, ISSN: 1050-2947.

Kim, Y. S.& Noz, M.E. (1991). *Phase space picture of Quantum Mechanics*. World Scientific, Singapore, pp. 37–55.

Kim, M. S.; Son, W. Buzek, V. & Knight, P. L. (2002). Entanglement by a beam splitter: Nonclassicality as a prerequisite for entanglement. *Physical Review A*, Vol. 65, article 032323, ISSN: 1050-2947.

Kirkwood, J. G. (1933). Quantum Statistics of Almost Classical Assemblies. *Physical Review* , Vol. 44, No. 1, pp. 31-37.

Klauder J. R. (1964) Continuous representation theory. II. Generalized relation between quantum and classical dynamics. *Journal of Mathematical Physics*, Vol. 5, No. 2, pp. 177-187. ISSN: 0022-2488.

Kogelnik, H. (1979). In *Integrated Optics*, ed. by Tamir, T. Springer Verlag, Berlin, pp. 15–81.

Kominis, Y. & Hizanidis, K. (2002). The Hamiltonian perturbation approach of two interacting nonlinear waves or solitary pulses in an optical coupler. *Physica D*, Vol. 173, No. 3, pp. 204-225, ISSN: 0167-2789.

Lauterborn, W.; Kurz, T. & Wiesenfeldt, W. T. (1993). *Coherent Optics*. Springer-Verlag, Berlin.

Lauterborn, W.; Kurz, T. & Parlitz, U. (1997). *Journal of the Franklin Institute*, Vol. 334, pp. 865.

Lee, H.W. (1995). Theory and application of the quantum phase-space distribution functions. *Physics Reports*, Vol. 259, No. 3, pp. 147-211, ISSN: 0370-1573.

Leonhardt U. (2001). *Measuring the Quantum State of Light*. Cambridge University Press, Cambridge.

Leontovich, M. & Fock, V. (1946). Solution of the problem of electromagnetic waves along the earth surface by the method of parabolic equations. *Soviet Journal of Physics*, Vol. 10, pp. 13-24.

Lesurf, J.C. et. al., (1993). Selected Papers on Gaussian Beam Mode Optics for Millimeter Wave and Terahertz Systems. S P I E - International Society for Optical Engineering, New York.

Lohmann A. (1980). The Wigner function and its optical production. *Optics Communications*, Vol. 32, pp. 32-38, ISSN: 0030-4018.

Lopez Viera J. C.; Rivera A. L.; Smirnov Yu. F. & Frank, A. (2002) Simple Evaluation of FranckâĂŞCondon Factors and Non-Condon Effects in the Morse Potential. *International Journal of Quantum Chemistry*, Vol. 88, No. 2, pp. 280-295, ISSN: 0020-7608.

Louisell, W. H. (1973). *Quantum Statistical Properties of Radiation*. Wiley, New York.

Lvovsky, A.I.; Hansen, H.; Aichele, T.; Benson, O.; Mlynek, J. & Schiller S. (2001) Quantum state reconstruction of the single-photon Fock state. *Physical Review Letters*, Vol. 87, Article 050402, ISSN 0031-9007.

Maanen, H. R. E. (1985). Duplication of the sampling frequency of periodically sampled signals for the calculation of the discrete Wigner distribution. *Journal of the Audio Engineering Society*. Vol. 33, pp. 892-901.

Marks, R. J. & Hall, M. W. (1979). Ambiguity function display using a single 1-D input. *Applied Optics*, Vol. 18, No. 15, pp. 2539-2540, ISSN: 1559-128X.

Mack, H. & Schleich, W. P. (2003). A photon viewed from Wigner phase space. *Optics and Photonics News* Vol. 14, pp. 28.

Manko, V. I. & Wolf, K.B. (1985). *The influence of aberrations in the optics of gaussian beam propagation*, Reporte de Investigación del departamento de Matemáticas, Universidad Autónoma Metropolitana, México.

Manko, V. I. (1986). Invariants and coherent states in fiber optics, In: *Lie Methods in Optics*, Ed. Sanchez Mondragon, J. & Wolf, K. B., pp. 193-206, Springer-Verlag, ISBN 3-540-16471-5, Berlin.

Marcuse, D. (1972). *Light transmission Optics*, Van Nostrand, New York.

Margenau, H. & Hill, R. N. (1961) Correlation between measurements in quantum theory. *Progress on Theoretical Physics* Vol. 26, pp. 722.

Mateeva Ts. & Sharlandjiev, P. (1986). Generation of a Wigner distribution function of complex signals by spatial filtering. *Optics Communications*, Vol. 57, pp. 153-158, ISSN: 0030-4018.

Mayer, J. E. & Band, W. (1947). On the Quantum Correction for Thermodynamic Equilibrium. *Journal of Chemical Physics*, Vol. 15, pp. 141-153.

Mecklenbrauker, W. et al. (1997). *The Wigner Distribution: Theory and Applications in Signal Processing*. Elsevier, Netherlands.

Mori, H.; Oppenheim, I. & Ross, J. (1962). Some topics in Quantum Statistics. The Wigner function and transport theory, In: *Studies in Statistical Mechanics* Vol. I, Ed. De Boer, J. & Uhlenbeck, G. E. North-Holland Pub., Amsterdam, pp. 213-298.

Moshinsky, M. & Smirnov, Y. F. (1996). *Harmonic Oscillator in Modern Physics: From Atoms to Quarks*, 2th ed., Gordon & Breach Publishing Group, New York.

Moyal, J. E. (1949). Quantum Mechanics as a statistical theory. *Proceedings of the Cambridge Philosophical Society*, Vol. 45, No. 1, pp. 99-124, ISSN: 0305-0041.

OConnell, R. F. (1983). The Wigner distribution function 50th birthday. *Foundations of Physics*. Vol.13, pp. 83-92, ISSN 0015-9018.

Oraevsky, A. N. (1998). Gaussian Beams and Optical Resonators. In *Proceedings of the Lebedev Physics Institute Seri3s* Vol. 222, Nova Science Publishers, New York.

Ou, Z. Y.; Pereira, S. F.; Kimble, H. J. & Peng, K.C. (1992). Realization of the EinsteinâĂŞPodolskyâĂŞRosen paradox for continuous variables. *Physical Review Letters*, Vol. 68, pp. 3663, ISSN 0031-9007.

Page, (1952). Instantaneous power spectra. *Journal of Applied Physics*, Vol.23, pp. 103.

Perinova, V.; Luis A. & Perina, J. (1998). *Phase in Optics*. World Scientific, Singapore.

Prasad, P. N. & Williams, D. J. (1991). *Introduction to Non-Linear Optical effects in Molecules and Polymers*. John Wiley & sons, New York.

Reyes, J.; Rodriguez, R.; Cotorogea, M. & Castano, V. M. (1999). *Optik*, Vol. 110, pp. 305, ISSN: 0030-4026.

Reyes, J.; Saenz, A.; Silva, A. & Castano, V. M. (1999). *Optik*, Vol. 110, pp. 527, ISSN: 0030-4026.

Reyes, J. & Castano, V. M. (2000). *Optik*, Vol. 111, pp. 219, ISSN: 0030-4026.

Rihaczek, A. W. (1968). Signal energy distribution in time and frequency. *IEEE Trans. Information Theory*, Vol. 14, pp. 369-376.

Rivera, A. L.; Chumakov, S. M.; Wolf, K. B. (1995). Hamiltonian foundation of geometrical anisotropic optics. *Journal of the Optical Society of America A*, Vol.12, No. 6, pp. 1380-1389, ISSN: 1084-7529.

Rivera, A. L.; Atakishiyev, N. M.; Chumakov, S. M.; Wolf, K. B. (1997). Evolution under polynomial Hamiltonians in quantum and optical phase spaces. *Physical Review A*, Vol. 55, No. 2, pp. 876-889, ISSN: 1050-2947.

Rivera, A. L. & Castano, V. M. (2010). Linear and non-linear symmetry properties of Gaussian wave packets. *Optik*, Vol. 121, No. 6, pp. 539-552, ISSN: 0030-4026.

Rivera, A. L. & Castano, V. M. (2010). Physical defects in fiber optics: A theoretical framework in phase space. *Optik*, Vol. 121, No. 17, pp. 1563-1569, ISSN: 0030-4026.

Royer, A. (1977). Wigner function as the expectation value of a parity operator. *Physical Review A*, Vol. 15, pp. 449-450, ISSN: 1050-2947.

Sako, T. & Diercksen, G. H. F. (2003). Confined quantum systems: spectral properties of the atoms helium and lithium in a power series potential. *Journal of Physics B*, Vol. 36, No. 7, pp. 1433-1458, ISSN 0953-4075.

Schleich, W. P. (2001). *Quantum Optics in Phase Space*. Wiley, Berlin.

Schempp, W. (1986). Analog radar signal design and digital signal processing: a Heisenberg nilpotent Lie group approach. In: *Lie Methods in Optics*, Ed. Sanchez Mondragon, J. & Wolf, K. B., (pp. 193-206), Springer-Verlag, ISBN 3-540-16471-5, Berlin.

Schrodinger, E. (1946). *Statistical Thermodynamics*. Cambridge University Press, Cambridge.

Sheppard, C. J. R. & Larkin, K.G. (2000). Focal shift, optical transfer function, and phase-space representations. *Journal of the Optical Society of America A*, Vol.17, No. 4, pp. 772-779, ISSN: 1084-7529.

Shimizu, M. et al. (1997). *Optical Fiber Amplifiers: Materials, Devices and Applications*. Artech House, New York.

Simon, M. K. (2002). *Probability Distributions Involving Gaussian Random Variables: A Handbook for Engineers and Scientists*. Kluwer Academic Publishers, New York.

Smith, T. B. (2006). Generalized Q-functions, *Journal of Physics A*, Vol. 39, No. 44, pp. 13747-13756, ISSN 1751-8113.

Smithey, D.T. Beck, M. & Raymer, M. G. (1993). Measurement of theWigner distribution and the density matrix of a light mode using optical homodyne tomography: application to squeezed states and the vacuum. *Physical Review Letters*, Vol. 70, pp. 1244-1256, ISSN 0031-9007.

Sohma, M. & Hirota, O. (2003). Capacity of a channel assisted by two-mode squeezed states. *Physical Review A*, Vol. 68, article 022303, ISSN: 1050-2947.

Soto, F. & Claverie, P. (1983). When is the Wigner function of multidimensional systems nonnegative? *Journal of Mathematical Physics*, Vol. 24, No. 1, pp. 97-100. ISSN: 0022-2488.

Stewart I. et al. (2002). *The Symmetry Perspective: From Equilibrium to Chaos in Phase Space and Physical Space*. Birkhauser, Boston.

Subotic, N. & Saleh, B. E. A. (1984). Generation of the Wigner distribution function of two-dimensional signals by a parallel optical processor. *Optics Letters*. Vol. 9, No. 471.

Sudarshan, E. C. G. (1963). Equivalence of Semiclassical and Quantum Mechanical Descriptions of Statistical Light Beams. *Physical Review Letters*, Vol. 10, pp. 277-279, ISSN 0031-9007.

Takabayasi, T. (1954). The Formulation of Quantum Mechanics in terms of Ensemble in Phase Space. *Progress on Theoretical Physics* Vol. 11, pp. 341.

Terletsky, Ya. P. (1937). *Z. Eksp. Teor. Fiz.* Vol. 7, pp. 1290.

Torchigin, V. P. & Torchigin A. V. (2003). An increase in the wavelength of light pulses propagating through a fiber. *Physics Letters A* , Vol. 311, No. 1, pp. 21-25, ISSN: 0375-9601.

von Neumann, J. (1927). *Gött. Nachr*, pp.273.

Voss, H. U.; Schwache, A.; Kurths, J. & Mitschke F. (1999). Equations of motion from chaotic data: A driven optical fiber ring resonator. *Physics Letters A*, Vol. 256, No. 1, pp. 47-54, ISSN: 0375-9601.

Walls, D. F. (1983). Squeezed states of light *Nature* Vol. 306, pp. 141-142, ISSN: 0028-0836.

Way, W.I. (1998). *Optical Fiber Communication*. SPIE-International Society for Optical Engineering, New York.

Weyl, (1927). *Z. Phys.*, Vol. 46, pp. 1.

Wigner, E. P. (1932). On the Quantum Correction For Thermodynamic Equilibrium. *Physical Review*, Vol. 40, No. 5, pp. 749-759.

Wigner, E. P. (1971). Quantum-mechanical distribution functions revisited. In *Perspectives in Quantum Theory*, Ed. Yourgrau, W. & van der Merwe, A. MIT Press, Cambridge.

Wolf, K. B. & Rivera, A. L. (1997). Holographic information in the Wigner function, *Optics Communications*, Vol. 144, No. 1-3, pp. 36-42, ISSN: 0030-4018.

Wolf, K. B. (2004). *Geometric Optics on Phase Space*. Springer, Berlin, ISBN 3-540-22039-9.

Woodward, P. W. *Probability and Information Theory with Applications to Radar*. Pergamon, Oxford.

Young, M. (2000). *Optics and Lasers: Including Fibers and Optical Waveguides*. Springer-Verlag, New York.

Yuen, H. P. (1976). Two-photon coherent states of the radiation field. *Physical Review A*, Vol. 13, pp. 2226, ISSN: 1050-2947.

Zalevsky, Z. & Mendlovic, D. (1997). Light propagation analysis in graded index fiber-Review and applications. *Fiberand Integrated Optics* Vol. 16, pp. 55-66, ISSN 0146-8030.

Permissions

The contributors of this book come from diverse backgrounds, making this book a truly international effort. This book will bring forth new frontiers with its revolutionizing research information and detailed analysis of the nascent developments around the world.

We would like to thank Dr Moh. Yasin, Professor Sulaiman W. Harun and Dr Hamzah Arof, for lending their expertise to make the book truly unique. They have played a crucial role in the development of this book. Without their invaluable contribution this book wouldn't have been possible. They have made vital efforts to compile up to date information on the varied aspects of this subject to make this book a valuable addition to the collection of many professionals and students.

This book was conceptualized with the vision of imparting up-to-date information and advanced data in this field. To ensure the same, a matchless editorial board was set up. Every individual on the board went through rigorous rounds of assessment to prove their worth. After which they invested a large part of their time researching and compiling the most relevant data for our readers. Conferences and sessions were held from time to time between the editorial board and the contributing authors to present the data in the most comprehensible form. The editorial team has worked tirelessly to provide valuable and valid information to help people across the globe.

Every chapter published in this book has been scrutinized by our experts. Their significance has been extensively debated. The topics covered herein carry significant findings which will fuel the growth of the discipline. They may even be implemented as practical applications or may be referred to as a beginning point for another development. Chapters in this book were first published by InTech; hereby published with permission under the Creative Commons Attribution License or equivalent.

The editorial board has been involved in producing this book since its inception. They have spent rigorous hours researching and exploring the diverse topics which have resulted in the successful publishing of this book. They have passed on their knowledge of decades through this book. To expedite this challenging task, the publisher supported the team at every step. A small team of assistant editors was also appointed to further simplify the editing procedure and attain best results for the readers.

Our editorial team has been hand-picked from every corner of the world. Their multi-ethnicity adds dynamic inputs to the discussions which result in innovative outcomes. These outcomes are then further discussed with the researchers and contributors who give their valuable feedback and opinion regarding the same. The feedback is then collaborated with the researches and they are edited in a comprehensive manner to aid the understanding of the subject.

Apart from the editorial board, the designing team has also invested a significant amount of their time in understanding the subject and creating the most relevant covers. They scrutinized every image to scout for the most suitable representation of the subject and create an appropriate cover for the book.

The publishing team has been involved in this book since its early stages. They were actively engaged in every process, be it collecting the data, connecting with the contributors or procuring relevant information. The team has been an ardent support to the editorial, designing and production team. Their endless efforts to recruit the best for this project, has resulted in the accomplishment of this book. They are a veteran in the field of academics and their pool of knowledge is as vast as their experience in printing. Their expertise and guidance has proved useful at every step. Their uncompromising quality standards have made this book an exceptional effort. Their encouragement from time to time has been an inspiration for everyone.

The publisher and the editorial board hope that this book will prove to be a valuable piece of knowledge for researchers, students, practitioners and scholars across the globe.

List of Contributors

Peter Horak and Francesco Poletti
University of Southampton, United Kingdom

Deng-Shan Wang
School of Science, Beijing Information Science and Technology University, Beijing, 100192, China

Edouard Brainis
Université libre de Bruxelles, Belgium

Arnaud Mussot and Alexandre Kudlinski
Laboratoire PhLAM, IRCICA, Université Lille 1, France

Shanglin Hou
School of Science, Lanzhou University of Technology, Lanzhou, China

Wei Qiu
Department of Physics, Liaoning University, Shenyang, China

Petr Drexler and Pavel Fiala
Department of Theoretical and Experimental Engineering, Brno University of Technology, Czech Republic

Hassan Abid Yasser
Thi-Qar University, Iraq Republic

Lynda Cherbi and Abderrahmane Bellil
Laboratory of Instrumentation LINS, University of Sciences and Technology Houary Boumedienne, USTHB, Algiers, Algeria

Ana Leonor Rivera
Centro de Fisica Aplicada y Tecnologia Avanzada, Universidad Nacional Autonoma de Mexico, Mexico

Printed in the USA
CPSIA information can be obtained
at www.ICGtesting.com
JSHW011359221024
72173JS00003B/352

9 781632 381491